The Probabilities of
Counting Codes

October 2011
Peter Müller

The author:
Peter Müller was born in 1962 in Heidelberg. After graduating as electrical engineer in 1992 at the University of Applied Sciences in Mannheim he started to work in the field of functional safety of computerized systems.
Already during the years of study iterative processes like calculating the Mandelbrot set attracted the author's interest.
Since 2002 he has been associated with *exida.com*, a leading company for functional safety.

Other publications:

Solving equations – using modified Fibonacci sequences
ISBN: 978-3-8423-3962-0

Bibliografische Information Der Deutschen Bibliothek
Die Deutsche Bibliothek verzeichnet diese Publikation in der Deutschen
Nationalbibliografie; detaillierte bibliografische Daten sind im Internet über
http://dnb.ddb.de abrufbar

1. Auflage 2011.10.01, Version 3

Herstellung und Verlag: Books on Demand GmbH,
Norderstedt

© 2011 Peter Müller

e-mail: pmueller@gmxpro.de

ISBN: 978-3-8423-8038-7

for you, Angie

Abstract

In many cases counters are used to count special events within an endless sequence of events. This paper discusses a specific calculation, related to the probability of a counter overrun if the special event occurs not in a predictable way but with a certain probability.
Different counter codes are compared with each other.

A probability formula is developed for special scenarios which are normally analyzed by state diagrams and which can be numerically solved by the related Markov chains.
The target is to enable a non-numerical discussion of the topic.

It is shown how formulas can be found based on the rules of the probability theory, and their correctness is verified by a comparison with Markov chains.

Table of content

Table of figures

Table of tables

1 Summary

It is shown how probabilities can be calculated for a certain experiment which makes use of an endless sequence of events. The outcome of the experiment depends on the current event and the previous events.

Sometimes it is interesting to look at the probabilities of special events to happen. Actually, the event itself is not necessarily the subject of interest, but it is the reaction to it. E.g. the dice roll result "6" is only from interest, if this result triggers a positive (or negative) reaction.
By evaluating the single events a reaction will be triggered with a probability P_{rt} or will not be triggered with the probability P_{nrt}.

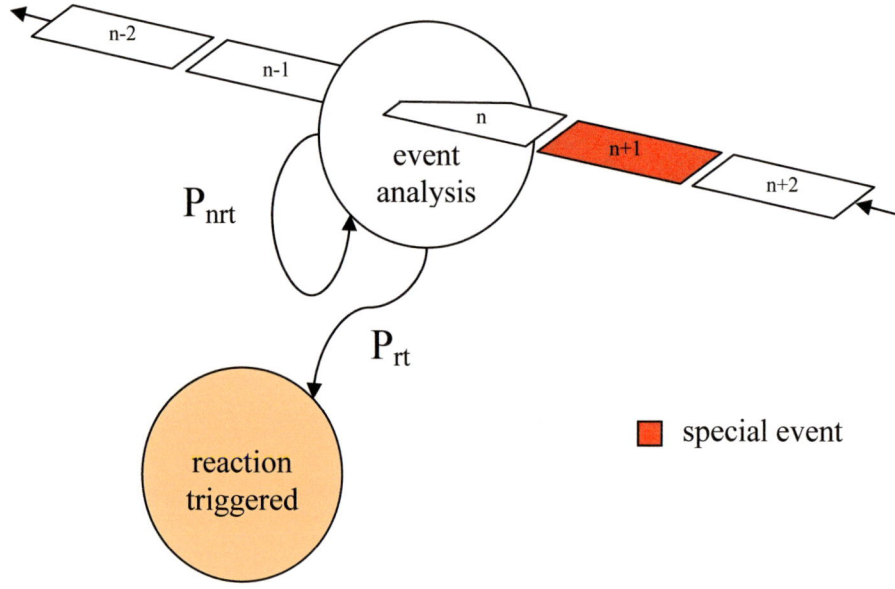

special event

In many scenarios it is not a single event that triggers a reaction, but several special events must occur within a defined number of events.

This is called "MooN", if **M** special events **out of N** sequential events occur the reaction will be triggered.

Such logic can be realized by using counters. Each event will be analyzed if it is a special event or not. If it is a special event it will be counted.

We call P_c the probability of the special event (the "**c**ounting" event) and P_{nc} the probability of the **n**on-counting event. As each event is either a counting- or a non-counting event we get $P_c + P_{nc} = 1$.

Depending on the used MooN code each event will trigger or not trigger the reaction. The probability P_{nrt} for the different MooN codes can be expressed by:

$$P_{nrt}^{N} - P_{nrt}^{N-1}P_{nc} - \sum_{i=0}^{M-2}\binom{N - M + i}{i}P_{nrt}^{M-2-i}P_{c}^{i+1}P_{nc}^{N-M+1} = 0$$

P_{nrt} is found as the value solving this equation in the range [0…1]. P_{rt} can always be calculated by:

$$P_{rt} = 1 - P_{nrt}$$

as each event will either trigger the reaction or not.

Based on the formula for the MooN codes the following table shows some results:

2002	M=2	N=2	$P_{nrt}^2 - P_{nrt}P_c P_{nc} - P_c P_{nc} = 0$
2003	M=2	N=3	$P_{nrt}^3 - P_{nrt}^2 P_c P_{nc} - P_c P_{nc}^2 = 0$
2004	M=2	N=4	$P_{nrt}^4 - P_{nrt}^3 P_c P_{nc} - P_c P_{nc}^3 = 0$
3003	M=3	N=3	$P_{nrt}^3 - P_{nrt}^2 P_c P_{nc} - P_{nrt}P_c P_{nc}P_{nc} - P_c^2 P_{nc} = 0$
3004	M=3	N=4	$P_{nrt}^4 - P_{nrt}^3 P_c P_{nc} - P_{nrt}P_c P_{nc}P_{nc}^2 - 2P_c^2 P_{nc}^2 = 0$
3005	M=3	N=5	$P_{nrt}^5 - P_{nrt}^4 P_c P_{nc} - P_{nrt}P_c P_{nc}P_{nc}^3 - 3P_c^2 P_{nc}^3 = 0$
3006	M=3	N=6	$P_{nrt}^6 - P_{nrt}^5 P_c P_{nc} - P_{nrt}P_c P_{nc}P_{nc}^4 - 4P_c^2 P_{nc}^4 = 0$
6009	M=6	N=9	$P_{nrt}^9 - P_{nrt}^8 P_c P_{nc} - P_{nrt}^4 P_c P_{nc}^4 - 4P_{nrt}^3 P_c^2 P_{nc}^4 - 10P_{nrt}^2 P_c^3 P_{nc}^4 - 20P_{nrt}P_c^4 P_{nc}^4 - 35P_c^5 P_{nc}^4 = 0$

Those formulas make it possible to compare the probabilities of the different counter codes with each other.

As an example it is shown that the probability of a 2ooN code for not triggering the reaction is by factor x higher than using the 1oo1 code:

$$\frac{P_{nrt\_2ooN}}{P_{nrt\_1oo1}} = x = 1 + \frac{P_c}{P_{nc}} \frac{1}{x^{N-1}}$$

$$x^N - x^{N-1} - \frac{P_c}{P_{nc}} = 0$$

P_c	P_{nc}	X				
		2oo2	2oo3	2oo4	2oo5	2oo6
0,999	0,001	32,110916	10,341360	5,889712	4,202483	3,353932
0,99	0,01	10,462430	4,984564	3,437433	2,744629	2,358016
0,90	0,10	3,541381	2,472368	2,047897	1,820096	1,677511
0,80	0,20	2,561553	2,000000	1,748403	1,604114	1,509828
0,70	0,30	2,107275	1,756380	1,585469	1,482742	1,413509
0,60	0,40	1,822876	1,591909	1,471129	1,395511	1,343145
0,50	0,50	1,618034	1,465571	1,380278	1,324718	1,285199
0,40	0,60	1,457427	1,360286	1,302028	1,262452	1,233480
0,30	0,70	1,323754	1,266982	1,230197	1,203968	1,184107
0,20	0,80	1,207107	1,179652	1,160116	1,145299	1,133568
0,10	0,90	1,100925	1,093006	1,086605	1,081283	1,076764
0,01	0,99	1,010001	1,009904	1,009809	1,009718	1,009628
0,001	0,999	1,001000	1,000999	1,000998	1,000997	1,000996

Looking at the numbers for $P_{nrt\_1oo1} = P_{nc} = 0{,}001$:
- using a 2oo2 code the probability that an event does not trigger the reaction is roughly 30 times higher
- using a 2oo3 code the probability that an event does not trigger the reaction is roughly 10 times higher

than using the 1oo1 code.

Interpreting the absolute number of P_{nrt} (see Annex 3.1) one can assume that the counting codes are meaningless if the probability of the special event is low, anyway (P_c is close to 0; P_{nc} is close to 1). The probability of not triggering the reaction is for all codes close to 1.
If the probability of the special event is high (P_c is close to 1; P_{nc} is close to 0) the counting codes will have effect related to avoidance of triggering the reaction.

Looking to the absolute numbers we get for $P_{nc} = 0{,}001$:

$$P_{nrt\_1oo1} = 0{,}001; \qquad P_{rt\_1oo1} = 0{,}999$$
$$P_{nrt\_2oo2} = 0{,}032; \qquad P_{rt\_2oo2} = 0{,}986$$
$$P_{nrt\_2oo3} = 0{,}010; \qquad P_{rt\_2oo3} = 0{,}990$$

The counting codes are effective, but the total probability of not triggering the reaction is low, which means that the absolute probability to trigger the reaction is high.

In situations where the probability of the special event is low and static, the usage of counting codes makes no big difference compared to the 1oo1 code, but they are very effective (in avoiding the reaction to be triggered) when the probability of the special event is getting temporarily (e.g. for only one or two events) high.

A typical scenario for what is discussed here can be found in industrial processes, where sequences of events are used to keep processes up and running. If single events are disturbed by e.g. environmental influences the process can get out of control. Therefore the events are checked for their status and are filtered out in case of a detected disturbance. If too many events are disturbed the process is shut down in a controlled manner.
The cause for the disturbances lies within the environmental conditions which cannot be stabilized. The changes of these environmental conditions might increase temporarily the probability of events to be disturbed.
The process itself defines how many events can be filtered out without negatively influencing the quality of the process.
If, for example, two "good" events in a sequence of five are needed, a 4oo5 counter code can be applied. Three detectably disturbed events can be tolerated within five, a fourth disturbed event will trigger the shutdown of the process. In other words, a short time environmental problem affecting three events will not cause a problem. A long time environmental problem will increase the probability of getting counting events and therefore it will trigger the shutdown of the process.

2   The rules of the game

As usual it all starts with a player throwing the dice.

For this paper this player is doing this continuously once every minute. He says that he is generating an "event" every minute.

The result of every die role is observed by another man who supervises for the numbers "2" and "4". If the observer saw one of those two numbers, he says that the "basic event" has happened.

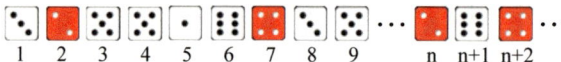

In the easiest version of the game, the observer commands the player to stop gambling, when he saw a single basic event. This rule is called "1oo1" (one out of one).

In the more complex version of the game the observer is waiting for a second basic event before he tells the player to stop the game. If the observer waits for N events this rule is called 2ooN (two out of N).

And of course this game exists in the MooN version where the observer waits for M basic events (to see a "2" or "4") within a sequence of N events (M out of N).

For example:

in the 2oo3 game, the observer would stop the game if he sees a "2" or "4" within three sequential events.

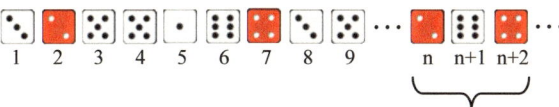

2 basic events within 3 events
→ stop gambling

To contribute to the game, the observer is counting the occurrences of the basic events.

The rule of the counting process is the following: whenever a "2" or a "4" is observed a counter is incremented. In case a "1", "3", "5" or "6" is observed the counter is decremented, but the counter must never go below zero.

2.1   Terminologies used within this paper

A counter will always be in a certain state. We will name the counter states in the following way. If the actual counter value is zero, we call this "0fc" (**fault count**[1] state 0), if it is one 1fc (**fault count** state 1), etc.

Any event will cause a transition of the counter state from one state to another. Those transitions are also identified by names:

There are events which increments the counter. If we increment the counter value to the state 1fc we call this

---

[1] The term „fault" seems to be strange at this moment. The reason is that **fault** counting is a typical application of what is discussed here.

1ci. If we increment the counter to the state 2fc we call this 2ci, etc.

Events that decrements the counter are accordingly called 1cd (**c**ounter **d**ecremented to the state 1fc), 2cd (counter decremented to the state 2fc), etc.

An event that does not change the counter value, which is only possible at the counter state 0fc, is called cz (**c**ounter **z**ero).

2.2 Basics for the probability calculation

The probabilities for the different numbers of a die are all 1/6.

The probability that the counter is incremented is called "P(c)" or shorter "P_c" - the **P**robability of a **c**ounting event;

the probability that the counter is decremented (or kept zero) is called "P(nc)" or shorter "P_{nc}" - the **P**robability of a **n**on-**c**ounting event.

As there are two options for the "basic event" which increments the counter, P_c equals 2/6 and as there are 4 options for the "non-basic events", which decrements the counter, P_{nc} equals 4/6.

As every event will somehow affect the counter we get:

$$P_c + P_{nc} = 1 \qquad (1)$$

"The counter either be incremented or decremented
(or remain zero) at all events"

The probability that the observer stop the game is called "P_{rt}" (**r**eaction **t**riggered) and accordingly the probability that the game is not stopped is s called "P_{nrt}" (**no reaction** triggered), and we get:

$$P_{rt} + P_{nrt} = 1 \qquad (2)$$

"The observer will either stop the game or not after each event"

The question that is discussed within this paper is: What is the probability of any event that the observer's reaction is triggered (based on a certain code evaluating the counter value).

The reaction can only be triggered by any event if it was not triggered already by the event before. Additionally the counter state after the last event is a pre-condition to trigger the reaction at the actual event.
In other words – we are talking about
"Conditional Probabilities"

As an example we look at the following code:
1. the counting events (P_c) increments a counter by 2,
2. the non-counting events (P_{nc}) decrements the counter by 1, but never be negative and
3. the reaction of the observer is triggered by a counter reading ≥ 3:

- the first event can result in
    - a counter reading of 0 (a non-counting event happened)
    - or 2 (a counting event occurred)

- the second event can result in
    - a counter reading of 4
    (if the counter reading was 2 before and
    the second event is a counting event)
    - a counter reading of 1
    (if the counter reading was 2 before and
    the second event is a non-counting event)
    - a counter reading of 2
    (if the counter reading was 0 before and
    the second event is a counting event)
    - a counter reading of 0
    (if the counter reading was 0 before and
    the second event is a non-counting event)

The figures 1 and 2 illustrates the counter treatment for
the first two events and for any event for the described
example.

In the Probability Theory the term of Conditional
Probability is defined:

$$P(A|B) = P(A \cap B)/P(B)$$

The meaning of this is: «What is the probability of the
event A to happen on the condition that event B has
happened.»
$P(A \cap B)$ represents the intersection of the events A and
B.
If we ask for example in Figure 2 for the probability that
the counter $= 1$ after event No. n+1.
Then there are two possibilities to ask:
- what is the probability to be in the state $1fc_{n+1}$, or

- what is the probability to get the transition $1ci_{n+1}$

These two questions are equivalent.

In most cases within this paper the second choice is used, to clearly identify which transition is investigated.

So the question:

What is the probability for the counter to be 1 after event No. n+1?

is answered by calculating the probability for the transition $1ci_{n+1}$ under the condition that the reaction was not triggered after event No. n (state: nrt_n):

$$P(1ci_{n+1}|\ nrt_n) = P(1ci_{n+1} \cap nrt_n)/P(nrt_n)$$

The intersection of $P(1ci_{n+1} \cap nrt_n)$ asks for the probability to get the transition 1ci by event No. n+1 based on the fact that after event No. n no reaction was triggered. This is only possible if the counter reading was 2 before (state $2fc_n$) and the second event was a non-counting event (P_{nc}) so the counter is decremented to 1:

$$P(1ci_{n+1} \cap nrt_n) = P(nc)P(2fc_n)$$

$P(1ci_{n+1}|\ nrt_n) = P(nc)P(2fc_n)\ /\ P(nrt_n)$
$P(1ci_{n+1}|\ nrt_n) = P(nc)P(2fc_n)\ /\ [P(2fc_n)+P(1fc_n)+P(0fc_n)]$

The complete example for the 2oo3 code is shown in Annex 2 for the first 3 events (including the check that $P(rt) + P(nrt) = 1$ after each event).

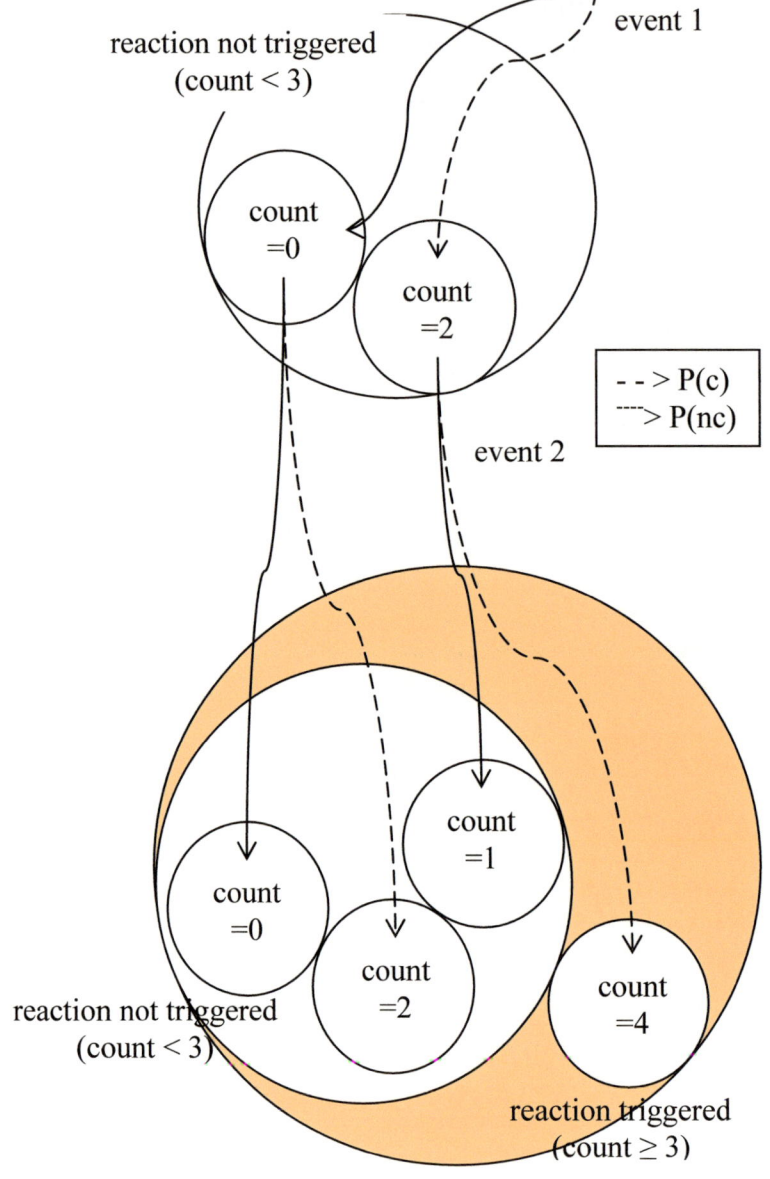

Figure 1: The first two events of the 2oo3 code

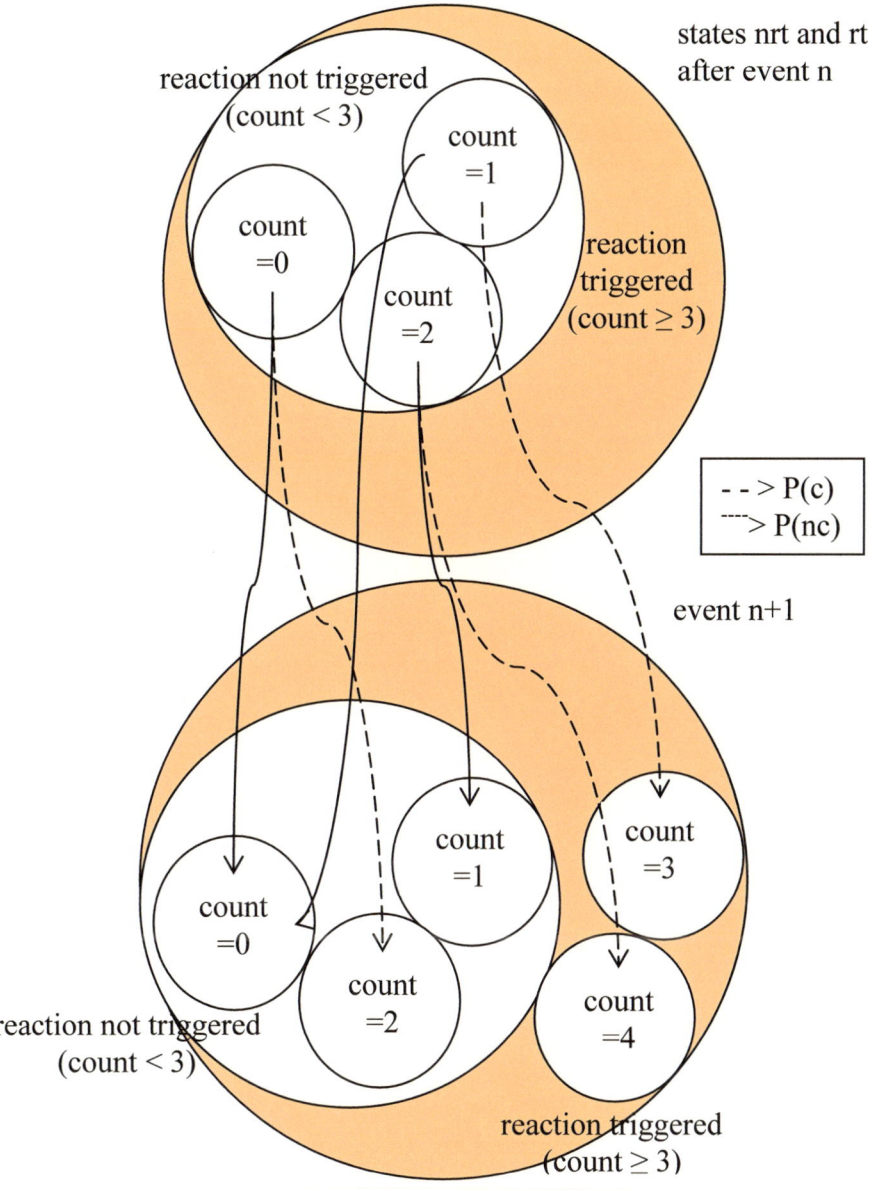

states nrt and rt
after event n

reaction not triggered
(count < 3)

count =1

count =0

count =2

reaction
triggered
(count ≥ 3)

- - > P(c)
···> P(nc)

event n+1

count =1

count =3

count =0

count =2

count =4

reaction not triggered
(count < 3)

reaction triggered
(count ≥ 3)

Figure 2: Any event of the 2oo3 code

2.3 A more abstract view

The Figure 2 can be drawn in a simplified way. Every event will be analyzed and will change the counter state (or the counter remains 0). It will increment or decrement the counter. The resulting counter reading will be evaluated and the reaction will be triggered or not:

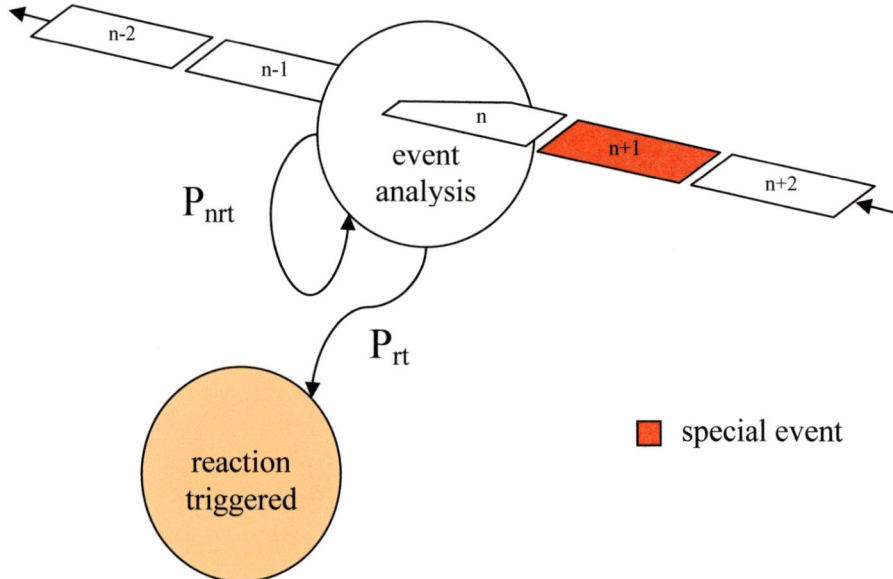

Figure 3: The principle view of the experiment

The target of this paper is to transform any counting code into the following simplified view. Any event will trigger / not trigger the reaction of the observer with the probabilities P(rt) and P(nrt).

The abstract view only shows the states that are reached after evaluating every single event:

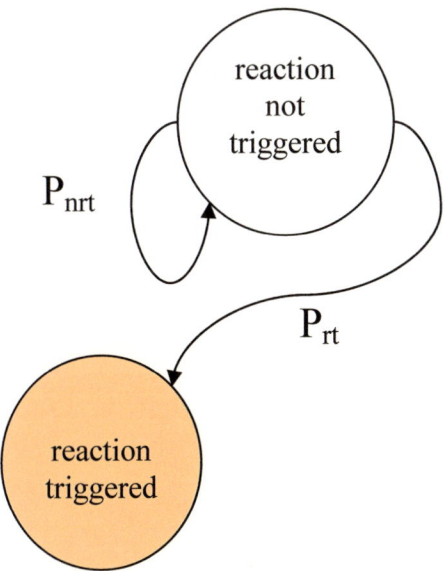

Figure 4: The simplified view for the result of any event

The following states of the game can be defined

- the reaction is not yet triggered | this state is called: | ntrig
  | and its probability is called: | P(ntrig)
- the reaction was triggered | this state is called: | trig
  | and its probability is called: | P(trig)

The following transitions can be defined

- the reaction is triggered | this transition is called: | rt
  | and its probability is called: | P(rt)
- the reaction is not triggered | this transition is called: | nrt
  | and its probability is called: | P(nrt)

3 The probabilities of the 1oo1 rule

A 1oo1 code would be, to increment a counter by 1 if a counting event occurs. A counter reading of 1 makes the observer to trigger the reaction.

The following transitions can be defined:

- the counter is incremented by 1 to 1   this transition is called: 1ci
  and its probability is called: P(1ci)
- the counter remains 0   this transition is called: cz
  and its probability is called: P(cz)

The following states of the counter can be defined

- the count is 0   this state is called: 0fc
  and its probability is called: P(0fc)
- the count is > 0   this state is called: 1fc
  and its probability is called: P(1fc)

The following transitions of the counter in the 1oo1 code are possible:

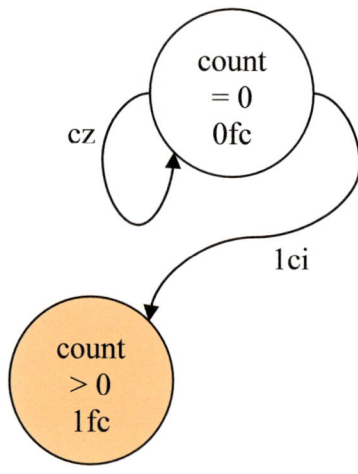

Figure 5: State transition diagram of the 1oo1 code

Only in case of the "red" marked state the reaction is triggered by the 1oo1 code.

3.1 The probability of no reaction

In Annex 1.1 the probabilities for P(rt) and P(nrt) are derived from the figure above. The reaction is:
- triggered if the counter reading is 1:

$$P(rt_n|nrt_{n-1}) \quad = P(1fc_n|nrt_{n-1}) = P_c$$

- not triggered if the counter reading is 0:

$$P(nrt_n|nrt_{n-1}) = P(0fc_n|nrt_{n-1}) = P_{nc}$$

The probability of any event to not trigger the reaction is given by:

$$P(nrt_\infty \mid nrt_{\infty-1}) = P(nc) = P_{nc}$$

3.2 The Markov Model of the 1oo1 code

A counting event P_c will always trigger the reaction of the observer. A non-counting event P_{nc} will never change the counter reading of zero.

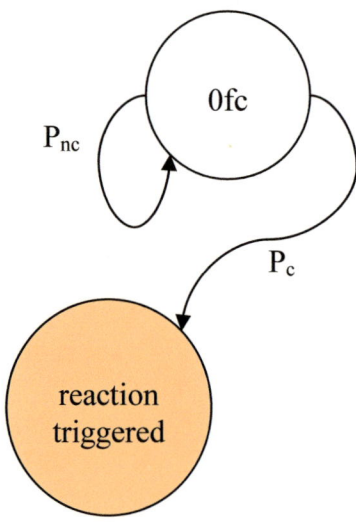

Figure 6: Markov Model of the 1oo1 code

The Markov transition matrix is defined by:

$$
\begin{array}{cc}
 & \begin{array}{cc} 0fc & triggered \end{array} \\
\begin{array}{c} 0fc \\ triggered \end{array} & \begin{array}{cc} P_{nc} & P_c \\ 0 & 1 \end{array}
\end{array}
\Rightarrow
\begin{pmatrix} P_{nc} & P_c \\ 0 & 1 \end{pmatrix}
$$

and the Markov start vector is:

$$
\begin{array}{cc}
0fc & triggered \\
\begin{pmatrix} 1 & \quad 0 \end{pmatrix}
\end{array}
$$

3.3 *Summary*

For the 1oo1 code there is a direct (linear) relation between the probabilities to trigger / not to trigger the reaction and the probability of counting events and non-counting events.

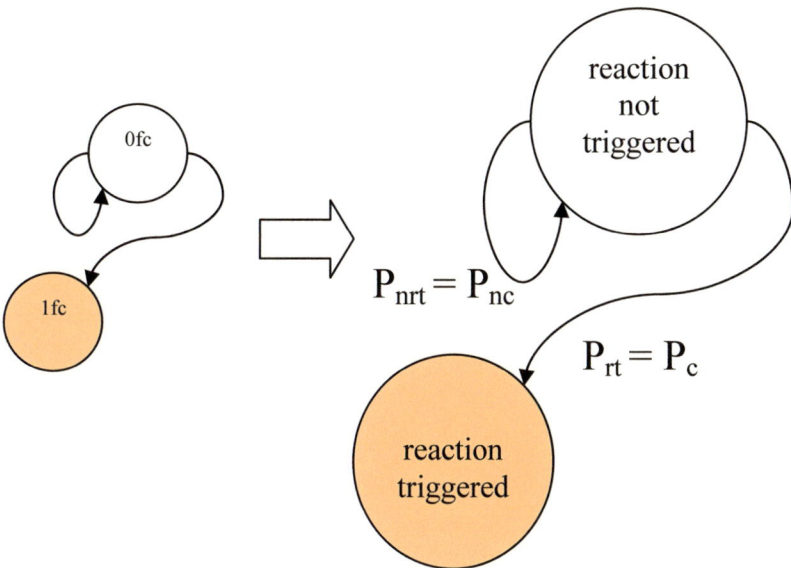

The probability of any event to not trigger / trigger the reaction is given by:

$$
\begin{aligned}
P_{nrt} &= P_{nc} \\
P_{rt} &= P_{c}
\end{aligned}
\qquad (3)
$$

4　The probabilities of the 2ooN rules

4.1　The probabilities of the 2oo2 rule

A 2oo2 code would be, to increment a counter by 1 when a counting event occurs, and to increment the counter again by 1 if the next event again is a counting event. A counter reading of 2 makes the observer to trigger the reaction.

When the counter reading is 1 and the next event is a non-counting event the counter is decremented to 0 again.

The following transitions can be defined:

- the counter is incremented by 1 to 1	this transition is called:	1ci
	and its probability is called:	P(1ci)
- the counter is incremented by 1 to 2	this transition is called:	2ci
	and its probability is called:	P(2ci)
- the counter is decremented by 1 to 0	this transition is called:	0cd
	and its probability is called:	P(0cd)
- the counter remains 0	this transition is called:	cz
	and its probability is called:	P(cz)

The following states of the counter can be defined

- the count is 0	this state is called:	0fc
	and its probability is called:	P(0fc)
- the count is 1	this state is called:	1fc
	and its probability is called:	P(1fc)
- the count is > 1	this state is called:	2fc
	and its probability is called:	P(2fc)

The following transitions of the counter in the 2oo2 code are possible:

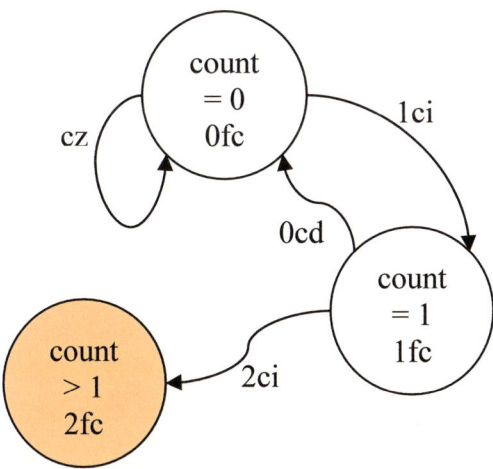

Figure 7: State transition diagram of the 2oo2 code

Only in case of the "red" marked state the reaction is triggered by the 2oo2 code.

4.1.1 The probability of no reaction

In Annex 1.2 the probabilities for P(rt) and P(nrt) are derived from the figure above. The reaction is:
- triggered when the counter reading is 2:

$P(rt_n|nrt_{n-1})$ $= P(2fc_n|nrt_{n-1})$

- not triggered when the counter reading is 0 or 1:

$P(nrt_n|nrt_{n-1}) = P(1fc_n|nrt_{n-1}) + P(0fc_n|nrt_{n-1})$

The probability of any event to not trigger the reaction is given by:

$$P(nrt_\infty \mid nrt_{\infty-1}) = P(nc) + \frac{P(c)P(nc)}{P(nrt_\infty \mid nrt_{\infty-1})}$$

$$P(nrt_\infty \mid nrt_{\infty-1})^2 - P(nc)P(nrt_\infty \mid nrt_{\infty-1}) - P(c)P(nc) = 0$$

4.1.2 The Markov Model of the 2oo2 code

All events that increments the counter are caused with the probability P(c):

$$P(1ci_n|nrt_{n-1}) + P(2ci_n|nrt_{n-1}) = P(c) = P_c$$

All events which decrements the counter or do not change it are caused with the probability P(nc):

$$P(0cd_n|nrt_{n-1}) + P(cz_n|nrt_{n-1}) = P(nc) = P_{nc}$$

As the different states are mutually exclusive (the counter can never show two different counter readings at the same time) the transitions are mutually exclusive. This means, if e.g. a non-counting event happens (P_{nc}) either the transition 0cd will happen ($P(0cd) = P_{nc}$); $P(cz) = 0$) or the transition cz will happen ($P(cz) = P_{nc}$); $P(0cd) = 0$). Therefore figure 7 can be modified to:

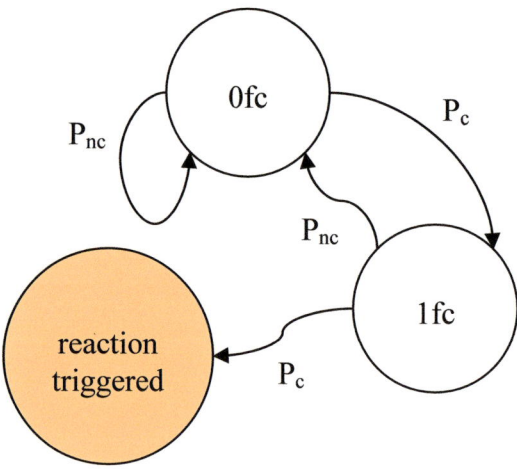

Figure 8: Markov Model of the 2oo2 code

The Markov transition matrix is defined by:

$$
\begin{array}{c c c c}
 & 0\,fc & 1\,fc & triggered \\
0\,fc & P_{nc} & P_c & 0 \\
1\,fc & P_{nc} & 0 & P_c \\
triggered & 0 & 0 & 1
\end{array}
\Rightarrow
\begin{pmatrix}
P_{nc} & P_c & 0 \\
P_{nc} & 0 & P_c \\
0 & 0 & 1
\end{pmatrix}
$$

and the Markov start vector is:

$$
\begin{array}{ccc}
0\,fc & 1\,fc & triggered \\
(\quad 1 & 0 & 0 \quad)
\end{array}
$$

4.1.3 Comparing the two approaches

To verify the correctness of the approaches, the results of the two approaches are compared as follows:
P_{nrt} as calculated according to chapter 4.1.1 is compared with the results of the Markov model.

To gain P_{nrt} from the Markov model the following calculation must be performed.
The Markov model calculation is an iterative process, where at first the start vector is multiplied with the transition matrix. In the next step the resulting vector is multiplied with the transition matrix, and so on. After each step the vector represents the probabilities of the different states.

For the example shown in chapter 2.2 with $P_c = 0,333...$
and $P_{nc} = 0,666...$ we get the following sequence:

step n	P(0fc)	P(1fc)	P(trig)
0	1,000	0,000	0,000
1	0,667	0,333	0,000
2	0,667	0,222	0,111
3	0,593	0,222	0,185
4	0,543	0,198	0,259
5	0,494	0,181	0,325
6	0,450	0,165	0,385
...	...	...	...
49	0,008	0,003	0,989
50	0,007	0,003	0,990

Table 1: Probabilities of the 2oo2 states (Markov)

P_{nrt} can be calculated by dividing the probability of not
being triggered P(ntrig) of step n by the one of step n-1.
For the above numbers we get with

$$P(ntrig) = 1 - P(trig) = P(0fc) + P(1fc)$$

step n	P(ntrig)	P(nrt) = P(ntrig)_n / P(ntrig)_{n-1}
0	1,000	-
1	1,000	1,000
2	0,889	0,889
3	0,815	0,917
4	0,741	0,909
5	0,675	0,911
6	0,615	0,911
...	...	...
49	0,011	0,911
50	0,010	0,911

Table 2: P(nrt) gained from Markov (2oo2)

From the Markov model of the 2oo2 code P_{nrt} can be determined to be 0,911 (for $P_c = 0,333...$ and $P_{nc} = 0,666...$)

P_{nrt} as calculated according to Annex 1.2 is based on the sequence of:

$$P(nrt_n|nrt_{n-1}) = P(nc) + P(c)P(nc) / P(nrt_{n-1})$$

P(nrt) after step 1 must be 1, as the 2oo2 code cannot trigger the reaction after the first message.

step n	P(nrt)
0	-
1	1,000
2	0,889
3	0,917
4	0,909
5	0,911
6	0,911
...	...
49	0,911
50	0,911

Table 3: P(nrt) from the 2oo2 equation

This shows that the P_{nrt} can either be determined by the Markov model or by solving the equation:

$$P(nrt_\infty \mid nrt_{\infty-1})^2 - P(nc)P(nrt_\infty \mid nrt_{\infty-1}) - P(c)P(nc) = 0$$

4.1.4 Summary

For the 2oo2 code there is a quadratic relation between the probabilities to trigger / not to trigger the reaction and the probabilities of counting / non-counting events.

For any event, it is always more likely that the 2oo2 code does not triggers the reaction than the 1oo1 code would do.

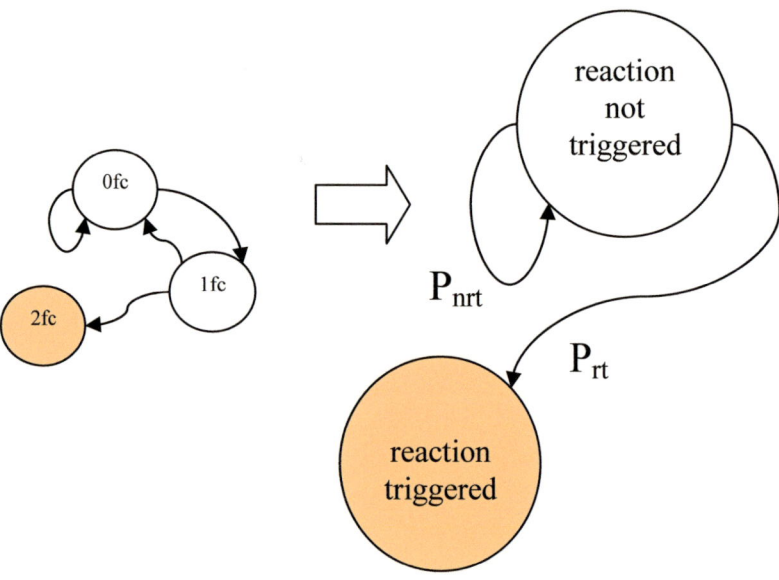

The probability of any event to not trigger / trigger the reaction is given by:

$$P_{nrt}^2 - P_{nc}P_{nrt} - P_cP_{nc} = 0 \qquad (4)$$
$$P_{rt} = 1 - P_{nrt}$$

4.2 The probabilities of the 2oo3 rule

A 2oo3 code would be, to increment a counter by 2 whenever a counting event occurs, and to decrement the counter by 1 if a non- counting event occurs (but never go below 0). A counter reading of ≥ 3 makes the observer to trigger the reaction.

This code triggers a reaction whenever at least two counting events in a sequence of three events are detected.

The following transitions can be defined:

- the counter is incremented by 2 to 2	this transition is called: and its probability is called:	2ci P(2ci)
- the counter is incremented by 2 to 3	this transition is called: and its probability is called:	3ci P(3ci)
- the counter is incremented by 2 to 4	this transition is called: and its probability is called:	4ci P(4ci)
- the counter is decremented by 1 to 1	this transition is called: and its probability is called:	1cd P(1cd)
- the counter is decremented by 1 to 0	this transition is called: and its probability is called:	0cd P(0cd)
- the counter remains 0	this transition is called: and its probability is called:	cz P(cz)

The following states of the counter can be defined

- the count is 0	this state is called: and its probability is called:	0fc P(0fc)
- the count is 1	this state is called: and its probability is called:	1fc P(1fc)
- the count is 2	this state is called: its probability is called:	2fcand P(2fc)
- the count is > 2	this state is called: and its probability is called:	3fc P(3fc)

The following transitions of the counter in the 2oo3 code are possible:

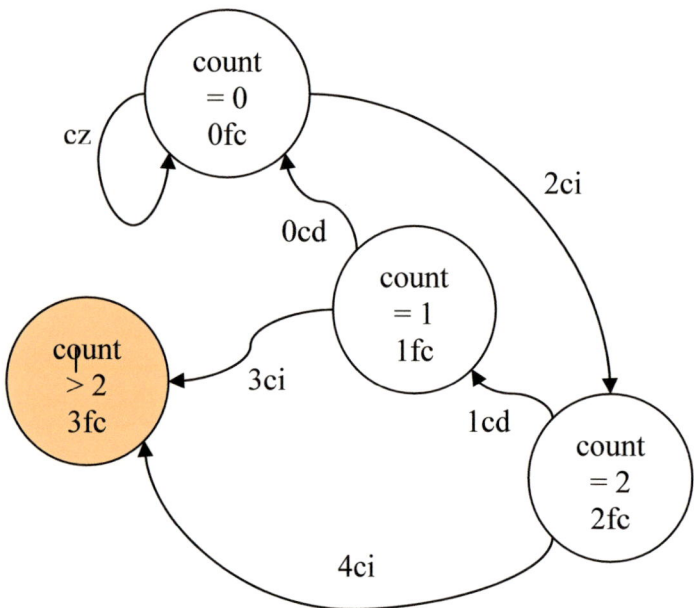

Figure 9: State transition diagram of the 2oo3 code

Only in case of the "red" marked state the reaction is triggered by the 2oo3 code.

4.2.1 The probability of no reaction

In Annex 1.3 the probabilities for P(rt) and P(nrt) are derived from the figure above. The reaction is:

- triggered when the counter reading is >2:

$$P(rt_n|nrt_{n-1}) = P(3fc_n|nrt_{n-1})$$

- not triggered when the counter reading is 0, 1 or 2:

$$P(nrt_n|nrt_{n-1}) = P(2fc_n|nrt_{n-1})+P(1fc_n|nrt_{n-1})+$$
$$P(0fc_n|nrt_{n-1})$$

The probability of any event to not trigger the reaction is given by:

$$P(nrt_\infty \mid nrt_{\infty-1}) = P(nc) + \frac{P(c)P(nc)^2}{P(nrt_\infty \mid nrt_{\infty-1})^2}$$

$$P(nrt_\infty \mid nrt_{\infty-1})^3 - P(nc)P(nrt_\infty \mid nrt_{\infty-1})^2 - P(c)P(nc)^2 = 0$$

4.2.2 The Markov Model of the 2oo3 code

With the same argumentation like in chapter 4.1.2 we get the Markov model of the 2oo3 code to:

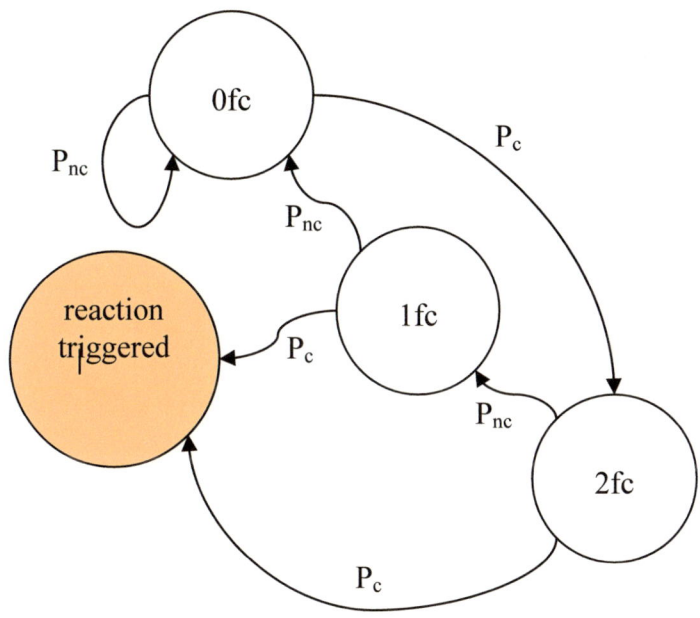

Figure 10: Markov Model of the 2oo3 code

The Markov transition matrix is defined by:

$$\begin{array}{c|cccc} & 0\,fc & 1\,fc & 2\,fc & triggered \\ \hline 0\,fc & P_{nc} & 0 & P_c & 0 \\ 1\,fc & P_{nc} & 0 & 0 & P_c \\ 2\,fc & 0 & P_{nc} & 0 & P_c \\ triggered & 0 & 0 & 0 & 1 \end{array} \Rightarrow \begin{pmatrix} P_{nc} & 0 & P_c & 0 \\ P_{nc} & 0 & 0 & P_c \\ 0 & P_{nc} & 0 & P_c \\ 0 & 0 & 0 & 1 \end{pmatrix}$$

and the Markov start vector is:

$$\begin{array}{cccc} 0\,fc & 1\,fc & 2\,fc & triggered \\ (1 & 0 & 0 & 0) \end{array}$$

4.2.3 Comparing the two approaches

Following the same approach as in chapter 4.1.3 it can be shown that the results of solving the equation or solving the Markov model are identical.

For the example shown in chapter 2.2 with $P_c = 0,333...$ and $P_{nc} = 0,666...$ we get the following sequence:

step n	P(0fc)	P(1fc)	P(2fc)	P(trig)
0	1,000	0,000	0,000	0,000
1	0,667	0,000	0,333	0,000
2	0,444	0,222	0,222	0,111
3	0,444	0,148	0,148	0,259
4	0,395	0,099	0,148	0,358
5	0,329	0,099	0,132	0,440
6	0,285	0,088	0,110	0,517
...	...	...		...
49	0,001	0,000	0,000	0,999
50	0,000	0,000	0,000	0,999

Table 4: Probabilities of the 2oo3 states (Markov)

P_{nrt} can be calculated by dividing the probability of not being triggered P(ntrig) of step n by the one of step n-1. For the above numbers we get with

$$P(ntrig) = 1 - P(trig) = P(0fc) + P(1fc)$$

step n	P(ntrig)	P(nrt) = $P(ntrig)_n / P(ntrig)_{n-1}$
0	1,0000000	-
1	1,0000000	1,000
2	0,8888889	0,889
3	0,7407407	0,833
4	0,6419753	0,867
5	0,5596708	0,872
6	0,4828532	0,863
...	...	...
49	0,0009340	0,865
50	0,0008077	0,865

Table 5: P(nrt) gained from Markov (2oo3)

From the Markov model of the 2oo2 code P_{nrt} can be determined to be 0,865 (for $P_c = 0,333...$ and $P_{nc} = 0,666...$)

P_{nrt} as calculated according to Annex 1.3 is based on the sequence of:

$$P(nrt_n|nrt_{n-1}) = P(nc) + P(c)P(nc)^2 / [P(nrt)_{n-1}P(nrt)_{n-2}]$$

P(nrt) for the steps 1 and 2 can be overtaken from the 2oo2 code, as for the first two steps there is no difference between 2oo2 and the 2oo3 code.

step n	P(nrt)
0	-
1	1,000
2	0,889
3	0,833
4	0,867
5	0,872
6	0,863
...	...
49	0,865
50	0,865

Table 6: P(nrt) from the 2oo3 equation

This shows that the P_{nrt} can either be determined by the Markov model or by solving the equation:

$$P(nrt_\infty \mid nrt_{\infty-1})^3 - P(nc)P(nrt_\infty \mid nrt_{\infty-1})^2 - P(c)P(nc)^2 = 0$$

4.2.4 Summary

For the 2oo3 code there is a cubic relation between the probabilities to trigger / not to trigger the reaction and the probabilities of counting / non-counting events.

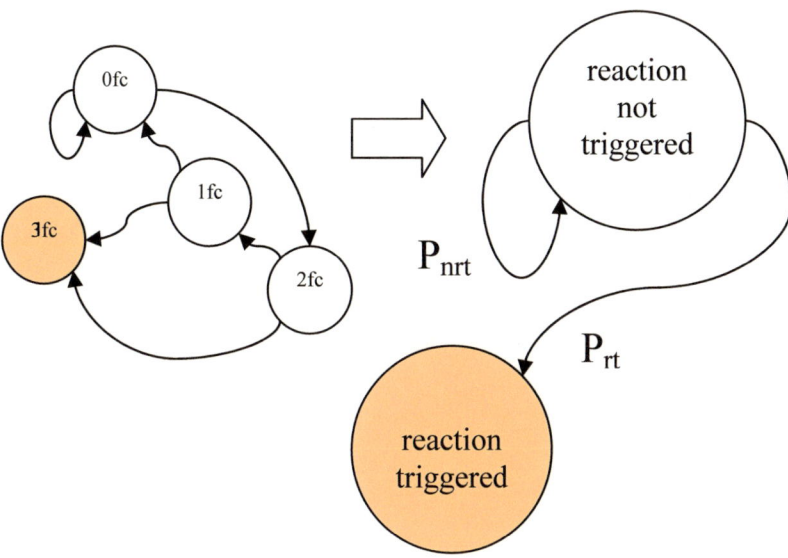

The probability of any event to not trigger / trigger the reaction is given by:

$$P_{nrt}^3 - P_{nc} P_{nrt}^2 - P_c P_{nc}^2 = 0 \qquad (5)$$
$$P_{rt} = 1 - P_{nrt}$$

4.3 The probabilities of the 2oo4 rule

A 2oo4 code would be, to increment a counter by 3 whenever a counting event occurs, and to decrement the counter by 1 if a non-counting event occurs (but never go below 0). A counter reading of >3 makes the observer to trigger the reaction.

This code triggers a reaction whenever there are counting events detected in two of four sequential events.

The following transitions can be defined:

- the counter is incremented by 3 to 3	this transition is called: and its probability is called:	3ci P(3ci)
- the counter is incremented by 3 to 4	this transition is called: and its probability is called:	4ci P(4ci)
- the counter is incremented by 3 to 5	this transition is called: and its probability is called:	5ci P(5ci)
- the counter is incremented by 3 to 6	this transition is called: and its probability is called:	6ci P(6ci)
- the counter is decremented by 1 to 2	this transition is called: and its probability is called:	2cd P(2cd)
- the counter is decremented by 1 to 1	this transition is called: and its probability is called:	1cd P(1cd)
- the counter is decremented by 1 to 0	this transition is called: and its probability is called:	0cd P(0cd)
- the counter remains 0	this transition is called: and its probability is called:	cz P(cz)

The following states of the counter can be defined

- the count is 0	this state is called: and its probability is called:	0fc P(0fc)
- the count is 1	this state is called: and its probability is called:	1fc P(1fc)
- the count is 2	this state is called: its probability is called:	2fcand P(2fc)
- the count is 3	this state is called: and its probability is called:	3fc P(3fc)
- the count is > 3	this state is called: and its probability is called:	4fc P(4fc)

The following transitions of the counter in the 2oo4 code are possible:

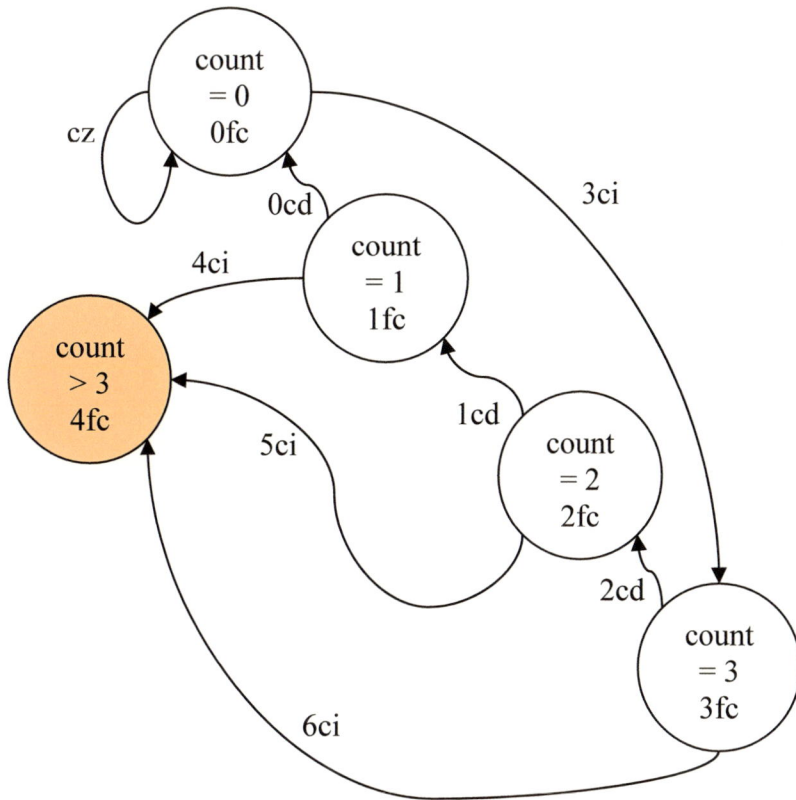

Figure 11: State transition diagram of the 2oo4 code

Only in case of the "red" marked state the reaction is triggered by the 2oo4 code.

4.3.1 The probability of no reaction

In Annex 1.4 the probabilities for P(rt) and P(nrt) are derived from the figure above. The reaction is:
- triggered when the counter reading is >3:

$P(rt_n|nrt_{n-1}) = P(4fc_n|nrt_{n-1})$

- not triggered when the counter reading is 0, 1, 2 or 3:

$P(nrt_n|nrt_{n-1}) = P(3fc_n|nrt_{n-1}) + P(2fc_n|nrt_{n-1}) +$
$$P(1fc_n|nrt_{n-1}) + P(0fc_n|nrt_{n-1})$$

The probability of any event to not trigger the reaction is given by:

$$P(nrt_\infty \mid nrt_{\infty-1}) = P(nc) + \frac{P(c)P(nc)^3}{P(nrt_\infty \mid nrt_{\infty-1})^3}$$

$$P(nrt_\infty \mid nrt_{\infty-1})^4 - P(nc)P(nrt_\infty \mid nrt_{\infty-1})^3 - P(c)P(nc)^3 = 0$$

4.3.2 The Markov Model of the 2oo4 code

With the same argumentation like in chapter 4.1.2 we get the Markov model of the 2oo4 code to:

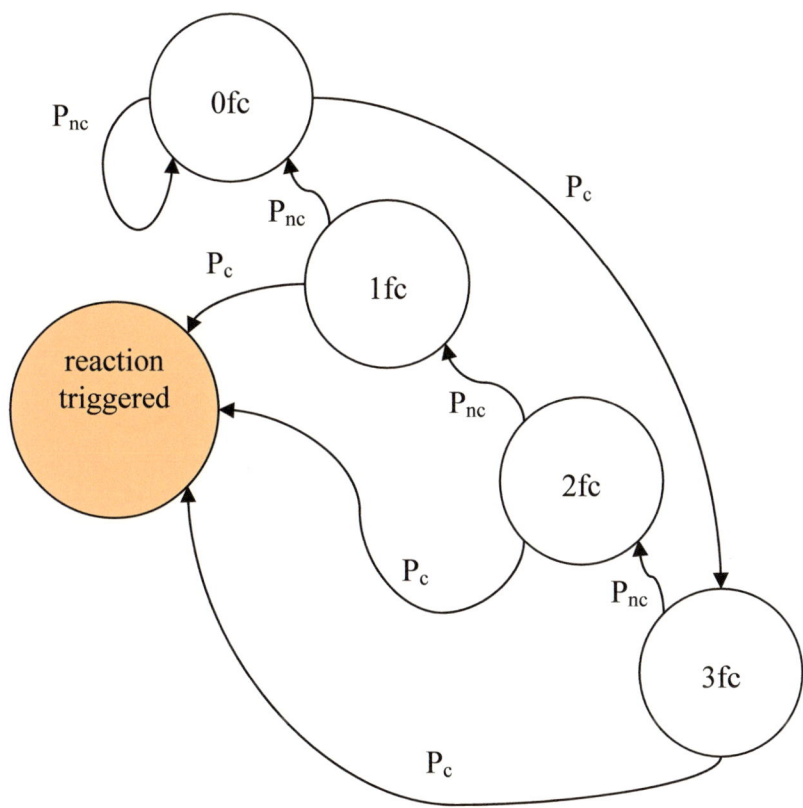

Figure 12: Markov Model of the 2oo4 code

The Markov transition matrix is defined by:

	$0\,fc$	$1\,fc$	$2\,fc$	$3\,fc$	triggered
$0\,fc$	P_{nc}	0	0	P_c	0
$1\,fc$	P_{nc}	0	0	0	P_c
$2\,fc$	0	P_{nc}	0	0	P_c
$3\,fc$	0	0	P_{nc}	0	P_c
triggered	0	0	0	0	1

$$\Rightarrow \begin{pmatrix} P_{nc} & 0 & 0 & P_c & 0 \\ P_{nc} & 0 & 0 & 0 & P_c \\ 0 & P_{nc} & 0 & 0 & P_c \\ 0 & 0 & P_{nc} & 0 & P_c \\ 0 & 0 & 0 & 0 & 1 \end{pmatrix}$$

and the Markov start vector is:

$0\,fc$	$1\,fc$	$2\,fc$	$3\,fc$	triggered
$(1$	0	0	0	$0)$

4.3.3 Comparing the two approaches

Following the same approach as in chapter 4.1.3 it can be shown that the results of solving the equation or solving the Markov model are identical.

For the example shown in chapter 2.2 with $P_c = 0,333...$
and $P_{nc} = 0,666...$ we get the following sequence:

step n	P(0fc)	P(1fc)	P(2fc)	P(3fc)	P(trig)
0	1,000	0,000	0,000	0,000	0,000
1	0,667	0,000	0,000	0,333	0,000
2	0,444	0,000	0,222	0,222	0,111
3	0,296	0,148	0,148	0,148	0,259
4	0,296	0,099	0,099	0,099	0,407
5	0,263	0,066	0,066	0,099	0,506
6	0,219	0,044	0,066	0,088	0,583
...	...	...	...		...
49	0,000	0,000	0,000	0,000	1,000
50	0,000	0,000	0,000	0,000	1,000

Table 7: Probabilities of the 2oo4 states (Markov)

P_{nrt} can be calculated by dividing the probability of not
being triggered P(ntrig) of step n by the one of step n-1.
For the above numbers we get with

$$P(ntrig) = 1 - P(trig) = P(0fc) + P(1fc)$$

step n	P(ntrig)	$P(nrt) =$ $P(ntrig)_n / P(ntrig)_{n-1}$
0	1,0000000	-
1	1,0000000	1,000
2	0,8888889	0,889
3	0,7407407	0,833
4	0,5925926	0,800
5	0,4938272	0,833
6	0,4170096	0,844
...	...	...
49	0,0001871	0,836
50	0,0001564	0,836

Table 8: P(nrt) gained from Markov (2oo4)

From the Markov model of the 2oo2 code P_{nrt} can be determined to be 0,836 (for $P_c = 0,333...$ and $P_{nc} = 0,666...$)

P_{nrt} as calculated according to Annex 1.4 is based on the sequence of:

$$P(nrt_n|nrt_{n-1})=P(nc)+P(c)P(nc)^3/[P(nrt)_{n-1}P(nrt)_{n-2}P(nrt)_{n-3}]$$

$P(nrt)$ for the steps 1, 2 and 3 can be overtaken from the 2oo3 code, as for the first three steps there is no difference between 2oo3 and the 2oo4 code.

step n	P(nrt)
0	-
1	1,000
2	0,889
3	0,833
4	0,800
5	0,833
6	0,844
...	...
49	0,836
50	0,836

Table 9: P(nrt) from the 2oo4 equation

This shows that the P_{nrt} can either be determined by the Markov model or by solving the equation:

$$P(nrt_\infty | nrt_{\infty-1})^4 - P(nc)P(nrt_\infty | nrt_{\infty-1})^3 - P(c)P(nc)^3 = 0$$

4.3.4 Summary

For the 2oo4 voting there is quartic relation between the probabilities to trigger / not to trigger the reaction and the probabilities of counting / non-counting events.

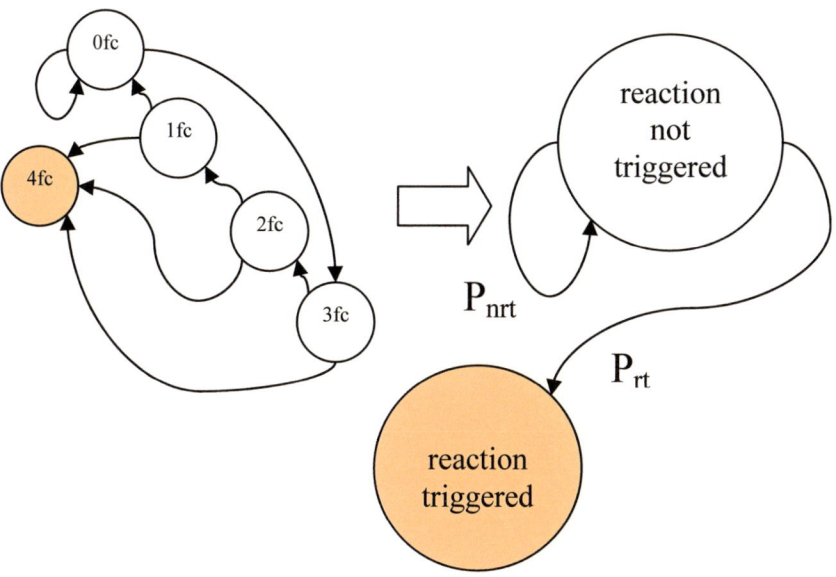

The probability of any event to not trigger / trigger the reaction is given by:

$$P_{nrt}^4 - P_{nc} P_{nrt}^3 - P_c P_{nc}^3 = 0 \qquad (6)$$
$$P_{rt} = 1 - P_{nrt}$$

5 The probabilities of the MooN rules

In order to determine a generic formula for calculating P_{nrt} it is necessary to be aware of one special feature of the Markov model.

As example we look again to the different probabilities of the 2oo3 Markov model:

step n	P(0fc)	P(1fc)	P(2fc)	P(trig)
0	1,000	0,000	0,000	0,000
1	0,667	0,000	0,333	0,000
2	0,444	0,222	0,222	0,111
3	0,444	0,148	0,148	0,259
4	0,395	0,099	0,148	0,358
5	0,329	0,099	0,132	0,440
6	0,285	0,088	0,110	0,517
7	0,249	0,073	0,095	0,583
8	0,215	0,063	0,083	0,639
9	0,185	0,055	0,072	0,688
10	0,160	0,048	0,062	0,730
11	0,139	0,041	0,053	0,767
12	0,120	0,036	0,046	0,798
13	0,104	0,031	0,040	0,825
14	0,090	0,027	0,035	0,849
15	0,078	0,023	0,030	0,869

Table 10: Probabilities of the 2oo3 states (Markov)

From the Markov model of the 2oo3 code P_{nrt} can be determined to be 0,865 (for $P_c = 0,333...$ and $P_{nc} = 0,666...$)

To get this result, we divide the probabilities P(xfc) of step n by the probability P(xfc) of the step n-1 before:

step n	$P(0fc)_n /$ $P(0fc)_{n-1}$	$P(1fc)_n /$ $P(1fc)_{n-1}$	$P(2fc)_n /$ $P(2fc)_{n-1}$
0	-	-	-
1	-	-	-
2	-	-	-
3	1,000	0,667	0,667
4	0,889	0,667	1,000
5	0,833	1,000	0,889
6	0,867	0,889	0,833
7	0,872	0,833	0,867
8	0,863	0,867	0,872
9	0,864	0,872	0,863
10	0,865	0,863	0,864
11	0,865	0,864	0,865
12	0,865	0,865	0,865
13	0,865	0,865	0,865
14	0,865	0,865	0,865
15	0,865	0,865	0,865

Table 11: P(nrt) gained from Markov (2oo3)

It can be seen that after some steps the model gets stable in its behavior and any

$$P(xfc)_n = P_{nrt} * P(xfc)_{n-1}.$$

and consequently

$$P(xfc)_n = P^2_{nrt} * P(xfc)_{n-2}$$

or in generic way

$$P(xfc)_n = P^y_{nrt} * P(xfc)_{n-y} \qquad (7)$$

5.1 The approach to get the generic formula

At the beginning of the game the counter reading is zero. In whatever state the counter is later, it can be exactly shown which sequence of counting / non-counting events has occurred.

E.g. if the 2oo3 counter is in the state 1fc, first a counting event P_c and then a non-counting event P_{nc} must have occurred in this sequence when the counter was in the state 0fc before.

In this way all counter states (except the "triggered state") can be expressed as a function of the 0fc state.

The idea is to express $P(0fc)$ only as a function of P_c, P_{nc}, P_{nrt} and former $P(0fc)_{n-x}$.

As $P(0fc)_{n-x}$ can be expressed by $P(0fc)_n / P^x_{nrt}$ (see (7)) it should be possible to get an expression for P_{nrt} as a function of P_c and P_{nc}.

$$P(0fc)_n = f(P_c, P_{nc}, P_{nrt}, P(0fc)_{n-x}) = f(P_c, P_{nc}, P_{nrt}, P(0fc)_n)$$

by canceling $P(0fc)_n$ we might get $1 = f(P_c, P_{nc}, P_{nrt})$

and therefore $P_{nrt} = f(P_c, P_{nc})$

5.2 Determining P(0fc) for any counter code

First, a generic Markov model is needed. We look to the already discussed pictures and develop the generic model based on the possible observations.

In the 1oo1 model (M =1, N = 1) one counting event is needed to trigger the reaction. Only the 0fc state and the triggered state exist.

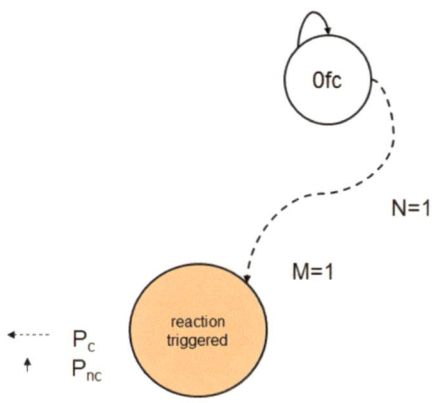

The following pictures are just intended to show how this model is changing when we increase M and N.

Two counting events P_c are needed to trigger the reaction in the 2ooN code.

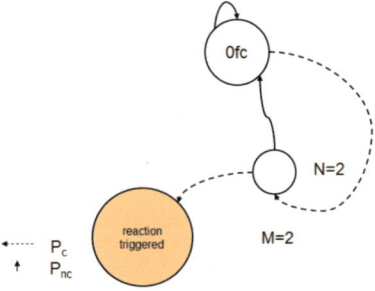

2oo2 (M = 2, N = 2) increasing N generates one new state

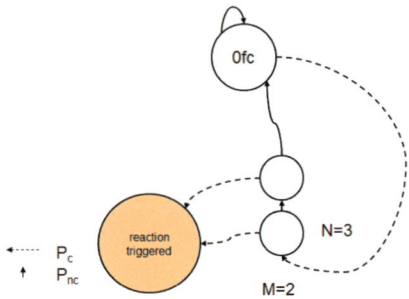

2oo3 (M = 2, N = 3) new states generates new rows

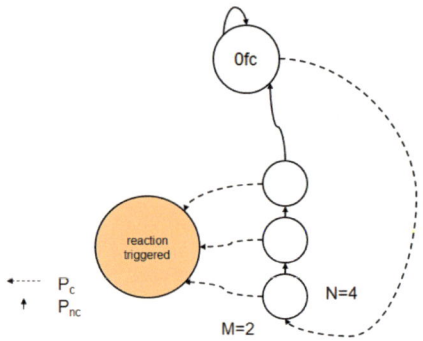

2oo4 (M = 2, N = 4) the number of state-rows depends on N

Three counting events P_c are needed to trigger the reaction in the 3ooN code.

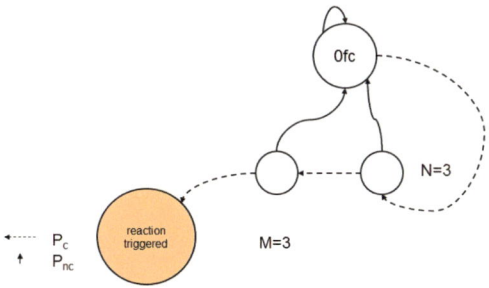

3oo3 (M = 3, N =3) another new state occurs as new column

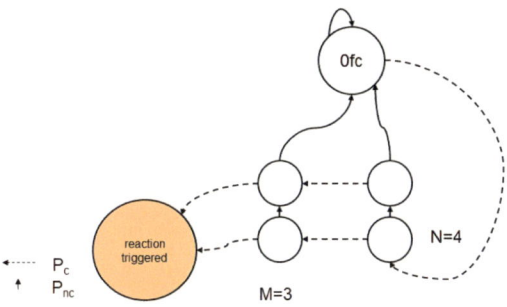

3oo4 (M = 3, N =4) like before, increasing N generates new rows

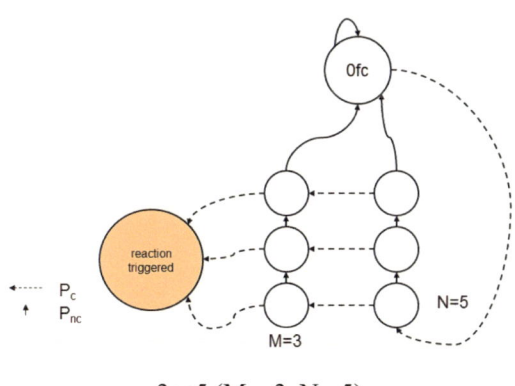

3oo5 (M = 3, N =5)

The above listed observations can be used to make a generic model (here shown for the 5oo9)

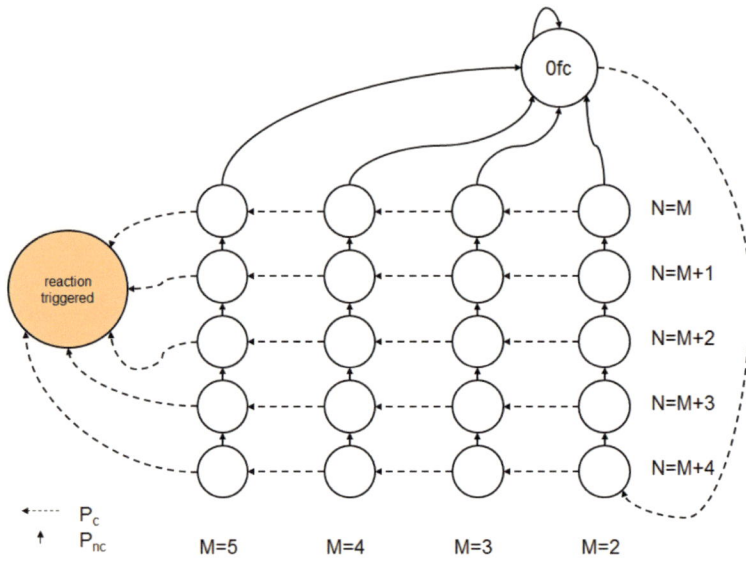

Figure 13: Markov Model of the 5oo9 code

Based on this generic model it is intended to develop a formula for P(0fc).

What can be seen is, that we can only be in the 0fc state after a non-counting event P_{nc} has occurred and we must have been in the states A, B, C, D or 0fc before:

$$P(0fc)_n = [P(0fc)_{n-1} + P(A)_{n-1} + P(B)_{n-1} + P(C)_{n-1} + P(D)_{n-1}] * P_{nc} \quad (8)$$

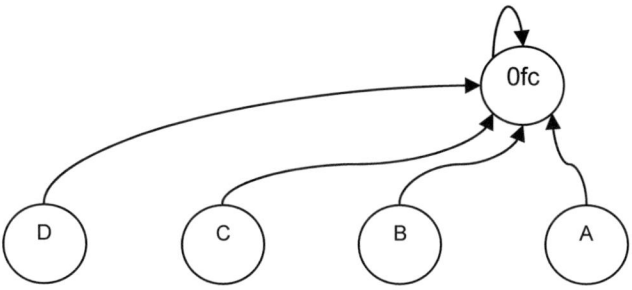

Figure 14: States that might lead to the state 0fc

Starting from the state 0fc in the above example, it requires:

trace 0fc	1 non-counting event to remain in the state 0fc
trace A	1 counting event and 4 non-counting events to come to the state A.
trace B	2 counting events and 4 non-counting events to come to the state B.
trace C	3 counting events and 4 non-counting events to come to the state C.
trace D	4 counting events and 4 non-counting events to come to the state D.

The trace 0fc is the simplest one. Being in the 0fc state and getting a non-counting event will not change the state.

For trace A there is only one sequence. After getting a counting event a sequence of non-counting events is required to reach again the 0fc state.
In the following figure this is indicated by marking each bubble with the number of chances to reach the state A. As there is only one way, the number is always 1.

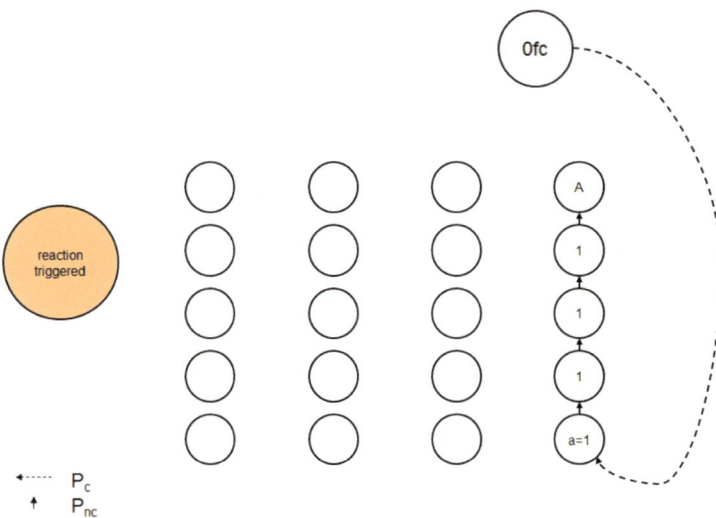

Figure 15: The probability to be in the state P(A)

5 events are needed to reach state A, starting from state 0fc.

$$P(A)_n = 1 * P(0fc)_{n-5} * P_c * P_{nc}^4$$

In this figure a=1, as that there is only 1 possible way to come to state A.

For $P(A)_{n-1}$ we get

$$P(A)_{n-1} = a * P(0fc)_{n-6} * P_c * P_{nc}^4$$

There are many different ways to come to the states B, C or D. This means that there are many different sequences of counting / non-counting events resulting in the different states.

For trace B, a first counting event is required but then there are several possibilities when the second counting event may occur.

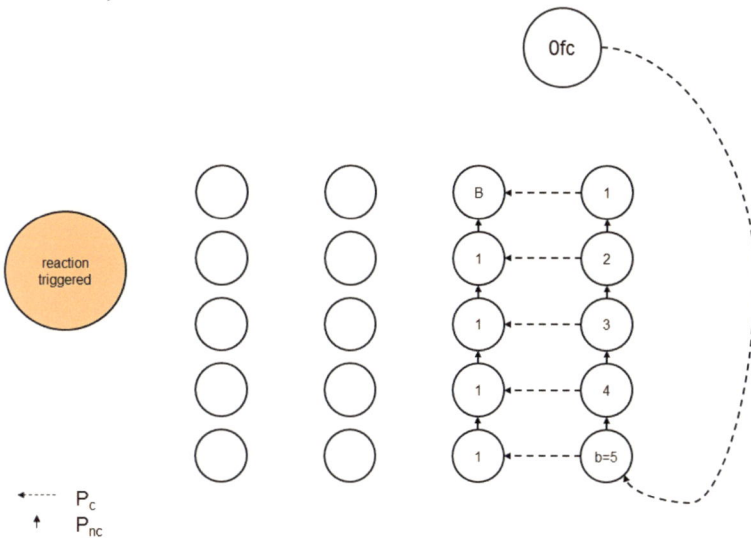

Figure 16: The probability to be in the state P(B)

6 events are needed to reach state B, starting from state 0fc.

$$P(B)_n = 5 * P(0fc)_{n-6} * P_c^2 * P_{nc}^4$$

In this figure b=5, as there are 5 possible ways to come to state B after the first counting event occurs.

For $P(B)_{n-1}$ we get

$$P(B)_{n-1} = b * P(0fc)_{n-7} * P_c^2 * P_{nc}^4$$

continuing this for trace C

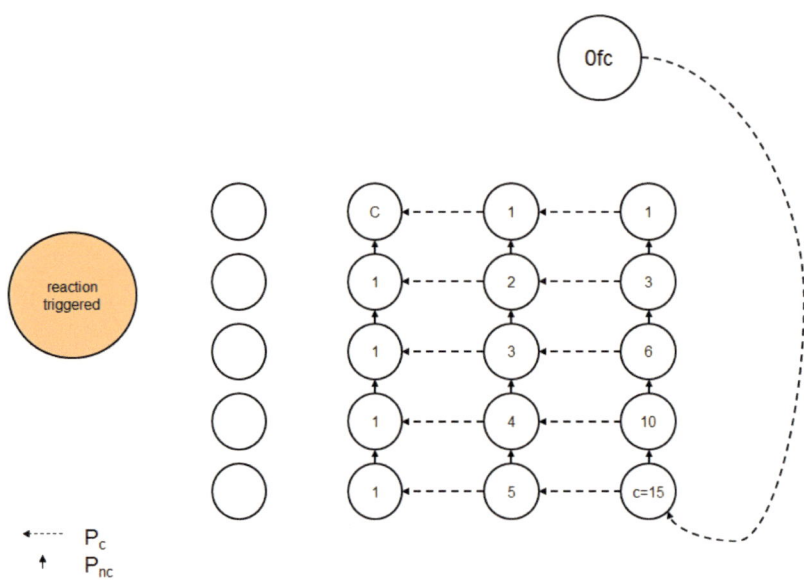

Figure 17: The probability to be in the state P(C)

7 events are needed to reach state C, starting from state 0fc.

$$P(C)_n = 15 * P(0fc)_{n-7} * P^3_c * P^4_{nc}$$

In this figure c=15, as there are 15 possible ways to come to state C after the first counting event occurs.

For $P(C)_{n-1}$ we get

$$P(C)_{n-1} = c * P(0fc)_{n-8} * P^3_c * P^4_{nc}$$

and finally for trace D

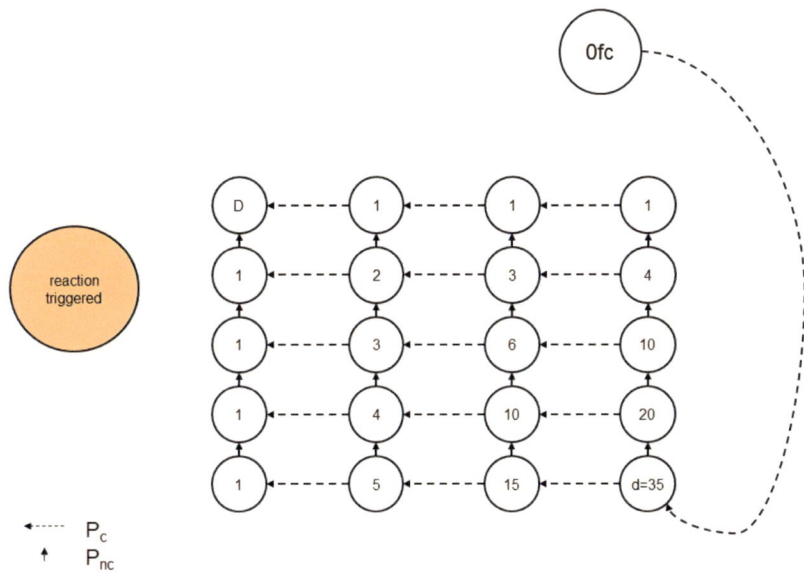

Figure 18: The probability to be in the state P(D)

8 events are needed to reach state D, starting from state 0fc.

$$P(D)_n = 35 * P(0fc)_{n-8} * P^4_c * P^4_{nc}$$

In this figure d=35, as there are 35 possible ways to come to state D after the first counting event occurs.

For $P(D)_{n-1}$ we get

$$P(D)_{n-1} = d * P(0fc)_{n-9} * P^4_c * P^4_{nc}$$

The factors a, b, c and d show the number of possible combinations of P_c and P_{nc} to reach the intended state. For the given example (M=5, N=9) these factors can be calculated as:

$$a = \binom{N-M+0}{0} = \binom{4}{0} = 1$$

$$b = \binom{N-M+1}{1} = \binom{5}{1} = 5$$

$$c = \binom{N-M+2}{2} = \binom{6}{2} = 15$$

$$d = \binom{N-M+3}{3} = \binom{7}{3} = 35$$

The indices within the formulas can also be expressed by N and M:

$$P(A)_{n-1} = a * P(0fc)_{n-N+M-2} * P_c^1 P_{nc}^{N-M}$$

$$P(B)_{n-1} = b * P(0fc)_{n-N+M-3} * P_c^2 P_{nc}^{N-M}$$

$$P(C)_{n-1} = c * P(0fc)_{n-N+M-4} * P_c^3 P_{nc}^{N-M}$$

$$P(D)_{n-1} = d * P(0fc)_{n-N+M-5} * P_c^4 P_{nc}^{N-M}$$

5.3 Determining the generic formula

Putting all the above formulas together we get:

$$P(0fc)_n = \left[P(0fc)_{n-1} + P(A)_{n-1} + P(B)_{n-1} + P(C)_{n-1} + P(D)_{n-1} \right] * P_{nc}$$

$$
\begin{aligned}
P(0fc)_n = P(0fc)_{n-1} P_{nc} &+ \binom{N-M+0}{0} P(0fc)_{n-N+M-2} P_c^1 P_{nc}^{N-M+1} + \\
&+ \binom{N-M+1}{1} P(0fc)_{n-N+M-3} P_c^2 P_{nc}^{N-M+1} + \\
&+ \binom{N-M+2}{2} P(0fc)_{n-N+M-4} P_c^3 P_{nc}^{N-M+1} + \\
&+ \binom{N-M+3}{3} P(0fc)_{n-N+M-5} P_c^4 P_{nc}^{N-M+1}
\end{aligned}
$$

using the relation (7): $P(0fc)_n = P_{nrt}^y * P(0fc)_{n-y}$

$$
\begin{aligned}
P(0fc)_n = \frac{P(0fc)_n}{P_{nrt}} P_{nc} &+ \binom{N-M+0}{0} \frac{P(0fc)_n}{P_{nrt}^{N-M+2}} P_c^1 P_{nc}^{N-M+1} + \\
&+ \binom{N-M+1}{1} \frac{P(0fc)_n}{P_{nrt}^{N-M+3}} P_c^2 P_{nc}^{N-M+1} + \\
&+ \binom{N-M+2}{2} \frac{P(0fc)_n}{P_{nrt}^{N-M+4}} P_c^3 P_{nc}^{N-M+1} + \\
&+ \binom{N-M+3}{3} \frac{P(0fc)_n}{P_{nrt}^{N-M+5}} P_c^4 P_{nc}^{N-M+1}
\end{aligned}
$$

by dividing by $P(0fc)_n$ we get

$$1 = \frac{1}{P_{nrt}} P_{nc} + \binom{N-M+0}{0} \frac{1}{P_{nrt}^{N-M+2}} P_c^1 P_{nc}^{N-M+1} +$$
$$\binom{N-M+1}{1} \frac{1}{P_{nrt}^{N-M+3}} P_c^2 P_{nc}^{N-M+1} +$$
$$\binom{N-M+2}{2} \frac{1}{P_{nrt}^{N-M+4}} P_c^3 P_{nc}^{N-M+1} +$$
$$\binom{N-M+3}{3} \frac{1}{P_{nrt}^{N-M+5}} P_c^4 P_{nc}^{N-M+1}$$

multiplying with P_{nrt}^N:

$$P_{nrt}^N = P_{nrt}^{N-1} P_{nc} + \binom{N-M+0}{0} P_{nrt}^{M-2} P_c^1 P_{nc}^{N-M+1} +$$
$$\binom{N-M+1}{1} P_{nrt}^{M-3} P_c^2 P_{nc}^{N-M+1} +$$
$$\binom{N-M+2}{2} P_{nrt}^{M-4} P_c^3 P_{nc}^{N-M+1} +$$
$$\binom{N-M+3}{3} P_{nrt}^{M-5} P_c^4 P_{nc}^{N-M+1}$$

this finally results in the generic formula:

$$P_{nrt}^N = P_{nrt}^{N-1} P_{nc} + \sum_{i=0}^{M-2} \binom{N-M+i}{i} P_{nrt}^{M-2-i} P_c^{i+1} P_{nc}^{N-M+1} \qquad (9)$$

2002	M=2	N=2	$P_{nrt}^2 - P_{nrt} P_{nc} - P_c P_{nc} = 0$
2003	M=2	N=3	$P_{nrt}^3 - P_{nrt}^2 P_{nc} - P_c P_{nc}^2 = 0$
2004	M=2	N=4	$P_{nrt}^4 - P_{nrt}^3 P_{nc} - P_c P_{nc}^3 = 0$
3003	M=3	N=3	$P_{nrt}^3 - P_{nrt}^2 P_{nc} - P_{nrt} P_c P_{nc} - P_c^2 P_{nc} = 0$
3004	M=3	N=4	$P_{nrt}^4 - P_{nrt}^3 P_{nc} - P_{nrt} P_c P_{nc}^2 - 2 P_c^2 P_{nc}^2 = 0$
3005	M=3	N=5	$P_{nrt}^5 - P_{nrt}^4 P_{nc} - P_{nrt} P_c P_{nc}^3 - 3 P_c^2 P_{nc}^3 = 0$
3006	M=3	N=6	$P_{nrt}^6 - P_{nrt}^5 P_{nc} - P_{nrt} P_c P_{nc}^4 - 4 P_c^2 P_{nc}^4 = 0$
6009	M=6	N=9	$P_{nrt}^9 - P_{nrt}^8 P_{nc} - P_{nrt}^4 P_c P_{nc}^4 - 4 P_{nrt}^3 P_c^2 P_{nc}^4 - 10 P_{nrt}^2 P_c^3 P_{nc}^4 - 20 P_{nrt} P_c^4 P_{nc}^4 - 35 P_c^5 P_{nc}^4 = 0$

Table 12: Several equations based on the generic formula

Analyzing formulas of this type, it can be shown that there is always one zero point within the range of [0...1]. Plotting those formulas we get two different types of graphs. The following figures are drawn for $P_{nc}=0,1$; $P_c=0,9$:

For even numbers N we always get curves like the following

For M=2, N=4 we get: $P_{nrt}^4 - 0,1 * P_{nrt}^3 - 0,0009 = 0$

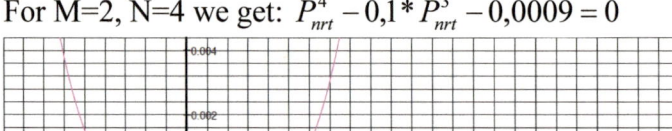

Figure 19: The 2oo4 graph

The zero point is at 0,2048

For M=5, N=8 we get:
$$P_{nrt}^8 - 0,1 * P_{nrt}^7 - 0,00009 * P_{nrt}^3 - 0,000324 * P_{nrt}^2 - 0,000729 P_{nrt}$$
$$-0,0013122 = 0$$

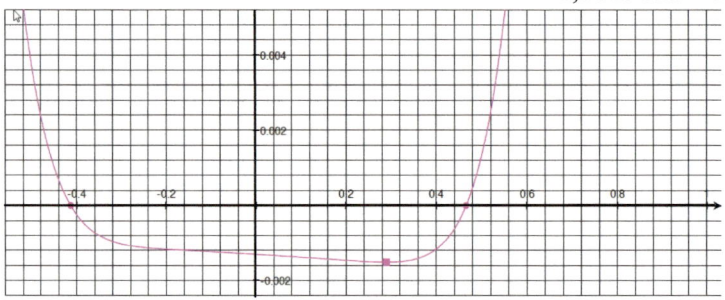

Figure 20: The 5oo8 graph

The zero point is at 0,4655

and for odd numbers N we always get curves like:

For M=4, N=7 we get:

$$P_{nrt}^7 - 0,1 * P_{nrt}^6 - 0,00009 * P_{nrt}^2 - 0,000324 * P_{nrt} - 0,000729 = 0$$

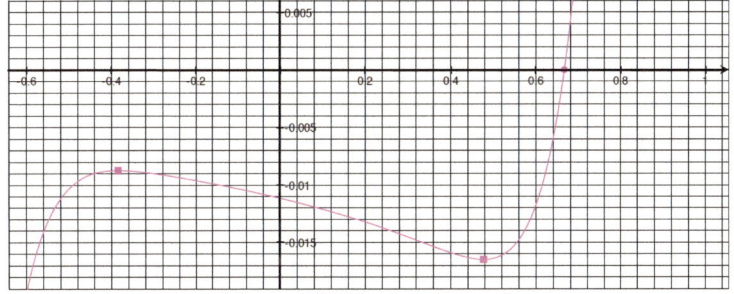

Figure 21: The 4o07 graph

The zero point is at 0,3814

For M=7, N=9 we get:

$$P_{nrt}^9 - 0,1 * P_{nrt}^8 - 0,0009 * P_{nrt}^5 - 0,00243 * P_{nrt}^4 - 0,004374 * P_{nrt}^3$$
$$-0,006561 * P_{nrt}^2 - 0,00885735 * P_{nrt} - 0,011160261 = 0$$

Figure 22: The 7o09 graph

The zero point is at 0,6658

It can be observed that there is always only one zero point within the range of [0…1] which represents the value of P_{nrt}.

5.4 A check of the generic formula

By taking the 6oo9 formula from the table above as example, it can be shown that the P(nrt) calculation is getting pretty convenient.

For the example shown in chapter 2 where $P_{nc} = 4/6$ and $P_c = 2/6$ the coefficients of the formula can be calculated:

$$P_{nc} = 0,6666667$$
$$P_c P_{nc}^4 = 0,0658436$$
$$4P_c^2 P_{nc}^4 = 0,0658436$$
$$10P_c^3 P_{nc}^4 = 0,0731596$$
$$20P_c^4 P_{nc}^4 = 0,0487731$$
$$35P_c^5 P_{nc}^4 = 0,0284509$$

Solving the equation we get P(nrt) = 0,989434

The Markov Model shows 22 different counter states. Therefore the Markov transition Matrix for the 6oo9 code has a size 22 x 22. After 50 iterations of the Matrix – Vector multiplication we get P(nrt) = 0,989431.

The both calculation principles come to the same result.

6 The difference of any 2ooN code and the 1oo1 code

Based on the formulas for the different codes, it should be possible to find a "factor", describing the difference between the 1oo1 code to any of the 2ooN codes.
To reach this, the formulas for the codes, as found before, will be interpreted:

1oo1: $P(nrt_n \mid nrt_{n-1}) = P(nc)$

2oo2: $P(nrt_n \mid nrt_{n-1}) = P(nc) + \dfrac{P(c)P(nc)}{P(nrt_n \mid nrt_{n-1})}$

2oo3: $P(nrt_n \mid nrt_{n-1}) = P(nc) + \dfrac{P(c)P(nc)^2}{P(nrt_n \mid nrt_{n-1})^2}$

2oo4: $P(nrt_n \mid nrt_{n-1}) = P(nc) + \dfrac{P(c)P(nc)^3}{P(nrt_n \mid nrt_{n-1})^3}$

As shown before, the value for $P(nrt_n \mid nrt_{n-1}) = P_{nrt}$ can be calculated for any 2ooN code by:

$$P_{nrt\_2ooN} = P(nc) + P(c)\dfrac{P(nc)^{N-1}}{P_{nrt\_2ooN}^{N-1}}$$

$$(P_{nrt\_2ooN}^N - P(nc)P_{nrt\_2ooN}^{N-1} - P(c)P(nc)^{N-1} = 0)$$

with $x = \dfrac{P_{nrt\_2ooN}}{P_{nrt\_1oo1}} = \dfrac{P_{nrt\_2ooN}}{P(nc)}$ we get:

$$x = 1 + \frac{P(c)}{P(nc)} \frac{1}{x^{N-1}} \qquad (10)$$

$$(x^N - x^{N-1} - \frac{P(c)}{P(nc)} = 0)$$

By finding the root for this equation, the ratio of $\dfrac{P_{nrt\_2ooN}}{P_{nrt\_1oo1}}$ can be determined for the different N.

This cannot be solved analytically for N > 4 anymore. A possible way to make a numeric approximation of possible solutions to such equations is shown in [D3].

6.1 Results

The following table shows the results for the different P(nrt) values of the 2ooN and the 1oo1 code (Annex 3.1 shows a more detailed table).

$P(nc)$	$P_{nrt\_2ooN} = P(nc) + P(c)\dfrac{P(nc)^{N-1}}{P_{nrt\_2ooN}^{N-1}}$				
P_{nrt_1oo1}	P_{nrt_2oo2}	P_{nrt_2oo3}	P_{nrt_2oo4}	P_{nrt_2oo5}	P_{nrt_2oo6}
0,01	0,104624	0,049846	0,034374	0,027446	0,023580
0,10	0,354138	0,247237	0,204790	0,182010	0,167751
0,20	0,512311	0,400000	0,349681	0,320823	0,301966
0,30	0,632183	0,526914	0,475641	0,444823	0,424053
0,40	0,729150	0,636764	0,588451	0,558204	0,537258
0,50	0,809017	0,732786	0,690139	0,662359	0,642600
0,60	0,874456	0,816172	0,781217	0,757471	0,740088
0,70	0,926628	0,886888	0,861138	0,842778	0,828875
0,80	0,965685	0,943722	0,928093	0,916240	0,906854
0,90	0,990833	0,983706	0,977944	0,973155	0,969087
0,99	0,999901	0,999805	0,999711	0,999621	0,999532
0,999	0,999999	0,999998	0,999997	0,999996	0,999995

Table 13: The probabilities of no reaction of different codes

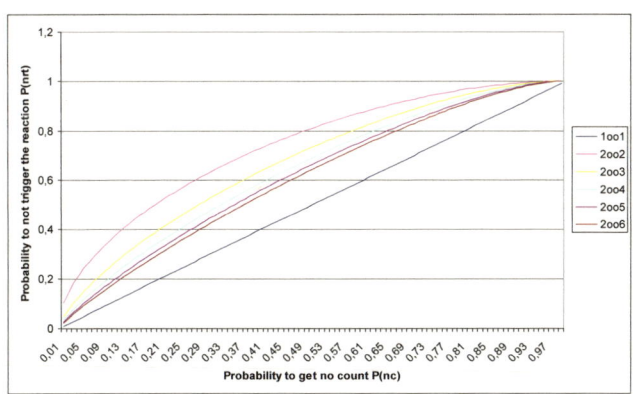

Figure 23: The probabilities of no reaction

The table above can be used to show the probabilities to trigger the reaction, and the influence of N can be interpreted:

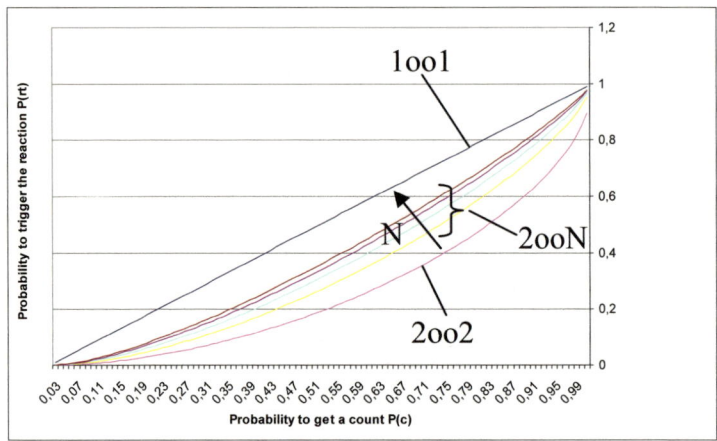

Figure 24: Interpreting the codes

- the 2ooN codes will trigger the reaction less likely the smaller N is. The 2oo2 code triggers least often the reaction.
- the 2oo2 code always has the lowest probability to trigger the reaction.
- the higher the number N of a 2ooN code is, the more identical to the 1oo1 code it is.

The following table and figure shows the factors between the 1oo1 code and the other codes (Annex 3.2 shows a more detailed table):

$P(nc)$	$x = 1 + \dfrac{P(c)}{P(nc)} \dfrac{1}{x^{N-1}}$				
	N=2	N=3	N=4	N=5	N=6
0,01	10,462430	4,984564	3,437433	2,744629	2,358016
0,10	3,541381	2,472368	2,047897	1,820096	1,677511
0,20	2,561553	2,000000	1,748403	1,604114	1,509828
0,30	2,107275	1,756380	1,585469	1,482742	1,413509
0,40	1,822876	1,591909	1,471129	1,395511	1,343145
0,50	1,618034	1,465571	1,380278	1,324718	1,285199
0,60	1,457427	1,360286	1,302028	1,262452	1,233480
0,70	1,323754	1,266982	1,230197	1,203968	1,184107
0,80	1,207107	1,179652	1,160116	1,145299	1,133568
0,90	1,100925	1,093006	1,086605	1,081283	1,076764
0,99	1,010001	1,009904	1,009809	1,009718	1,009628
0,999	1,001000	1,000999	1,000998	1,000997	1,000996

Table 14: Relation between the different codes (extract)

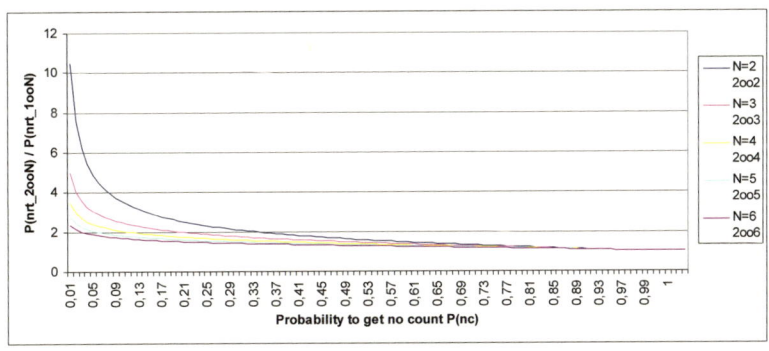

Figure 25: A comparison of the codes

-

This data shows that:
- if the probability to get no counter increment gets close to 1 (the probability to get a counter increment is low), all 2ooN codes will trigger the reaction with nearly the same probability.
- the 2ooN codes are most effective to avoid a reaction to be triggered if the probability to increment the counter is high. But the overall probability to trigger the reaction is high for all codes in that case.

7   References

[D1]	Taschenbuch der Mathematik Bronstein / Semendjajew / Musiol / Mühlig, 7te Auflage, 2008
[D2]	The Free Encyclopedia "Wikipedia"
[D3]	Solving equations – using modified Fibonacci sequences Peter Müller, 1te Auflage, 2010 ISBN: 978-3-8423-3962-0

Table 15: References

Annex 1

1.1 1oo1 Analysis

The following transitions of the counter in the 1oo1 code are possible:

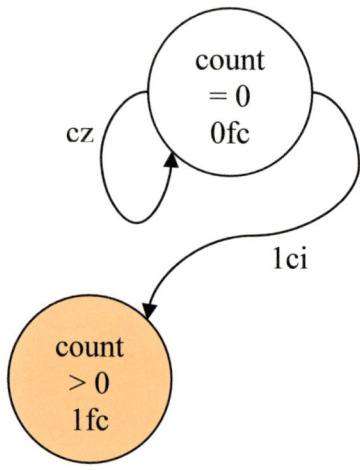

Only in case of the "red" marked state the reaction is triggered by the 1oo1 code.

1.1.1 Probabilities of the transitions and states for the 1oo1 code

The probabilities for the treatment (incrementing / not changing) of the counter, the counter states and finally the probabilities to trigger / not to trigger the reaction can be explained, based on the above picture:

1.1.2 Counter changes

The counter will:

- be incremented from 0 to 1, if the event number n is a counting event, with the probability $P(c)$, and the counter reading after event number n-1 was 0:

$$P(1ci_n|nrt_{n-1}) = P(1ci_n \cap nrt_{n-1}) / P(nrt_{n-1})$$
$$= P(c)P(0fc_{n-1}) / P(0fc_{n-1}) = P(c) = P_c$$

- remain 0, if event number n is a non-counting event, with the probability $P(nc)$, and the counter reading after event number n-1 was 0:

$$P(cz_n|nrt_{n-1}) = P(cz_n \cap nrt_{n-1}) / P(nrt_{n-1})$$
$$= P(nc)P(0fc_{n-1}) / P(0fc_{n-1}) = P(nc) = P_{nc}$$

1.1.3 Counter states

The counter reading can:
- only be 1 if the event 1ci occurred:

$$P(1fc_n|nrt_{n-1}) = P(1ci_n|nrt_{n-1}) = P_c$$

- only be 0 if the event cz occurred:

$$P(0fc_n|nrt_{n-1}) = P(cz_n|nrt_{n-1}) = P_{nc}$$

1.1.4 Reaction triggered

The reaction is:

- triggered if the counter reading is 1:

$$P(rt_n|nrt_{n-1}) = P(1fc_n|nrt_{n-1}) = P_c$$

- not triggered if the counter reading is 0:

$$P(nrt_n|nrt_{n-1}) = P(0fc_n|nrt_{n-1}) = P_{nc}$$

event number n	1	2	3
Reaction triggered by event n	$P(rt_1)$ $= P_c$	$P(rt_2 \mid nrt_1) = \dfrac{P(rt_2 \cap nrt_1)}{P(nrt_1)}$ $= \dfrac{P(1fc_2)P(0fc_1)}{P(0fc_1)} = \dfrac{P_c P_{nc}}{P_{nc}}$ $= P_c$	$P(rt_3 \mid nrt_2)$ $= P_c$
Reaction not triggered by event n	$P(nrt_1)$ $= P_{nc}$	$P(nrt_2 \mid nrt_1) = \dfrac{P(nrt_2 \cap nrt_1)}{P(nrt_1)}$ $= \dfrac{P(0fc_2)P(0fc_1)}{P(0fc_1)} = \dfrac{P_{nc} P_{nc}}{P_{nc}}$ $= P_{nc}$	$P(nrt_3 \mid nrt_2)$ $= P_{nc}$

Table 16: The probabilities of the 1oo1 code

1.2 2oo2 Analysis

The following transitions of the counter in the 2oo2 code are possible:

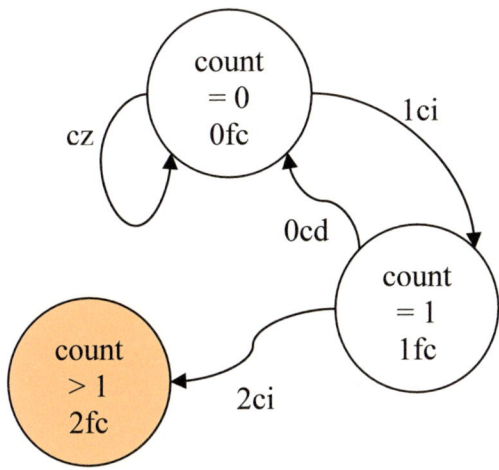

Only in case of the "red" marked state the reaction is triggered by the 2oo2 code.

1.2.1 Probabilities of the transitions and states for the 2oo2 code

The probabilities for the treatment (incrementing / decrementing / not changing) of the counter, the counter states and finally the probabilities to trigger / not to trigger the reaction can be explained, based on the above picture:

1.2.2 Counter changes

The counter will:

- be incremented from 0 to 1, if the event number n is a counting event, with the probability $P(c)$, and the counter reading after event number n-1 was 0:

$$P(1ci_n|nrt_{n-1}) = P(1ci_n \cap nrt_{n-1}) / P(nrt_{n-1})$$
$$= P(c)P(0fc_{n-1}) / (P(0fc_{n-1})+P(1fc_{n-1}))$$

- be incremented from 1 to 2, if the event number n is a counting event, with the probability $P(c)$, and the counter reading after event number n-1 was 1:

$$P(2ci_n|nrt_{n-1}) = P(2ci_n \cap nrt_{n-1}) / P(nrt_{n-1})$$
$$= P(c)P(1fc_{n-1}) / (P(0fc_{n-1})+P(1fc_{n-1}))$$

- be decremented from 1 to 0, if the event number n is a non-counting event, with the probability $P(nc)$, and the counter reading after event number n-1 was 1:

$$P(0cd_n|nrt_{n-1}) = P(0cd_n \cap nrt_{n-1}) / P(nrt_{n-1})$$
$$= P(nc)P(1fc_{n-1}) / (P(0fc_{n-1})+P(1fc_{n-1}))$$

- remain 0, if the event number n is a non-counting event, with the probability $P(nc)$, and the counter reading after event number n-1 was 0:

$$P(cz_n|nrt_{n-1}) = P(cz_n \cap nrt_{n-1}) / P(nrt_{n-1})$$
$$= P(nc)P(0fc_{n-1}) / (P(0fc_{n-1})+P(1fc_{n-1}))$$

1.2.3 Counter states

The counter reading can:
- only be 1 when the event 1ci occurred:
$$P(1fc_n|nrt_{n-1}) = P(1ci_n|nrt_{n-1})$$
- only be 2 when the event 2ci occurred:
$$P(2fc_n|nrt_{n-1}) = P(2ci_n|nrt_{n-1})$$
- be 0 when the event 0cd or the event cz occurred:
$$P(0fc_n|nrt_{n-1}) = P(0cd_n|nrt_{n-1}) + P(cz_n|nrt_{n-1})$$

1.2.4 Reaction triggered

The reaction is:
- triggered when the counter reading is 2:
$$P(rt_n|nrt_{n-1}) = P(2fc_n|nrt_{n-1})$$
- not triggered when the counter reading is 0 or 1:
$$P(nrt_n|nrt_{n-1}) = P(1fc_n|nrt_{n-1}) + P(0fc_n|nrt_{n-1})$$

1.2.5 Interpretations

The counter will be handled by one of the above actions for sure ($P_c + P_{nc} = 1$), this means that the sum of the above probabilities is the almost surely event:
$$P(1ci_n|nrt_{n-1})+P(2ci_n|nrt_{n-1})+P(0cd_n|nrt_{n-1})+P(cz_n|nrt_{n-1})$$
$$= 1$$
All events that increment the counter are caused with the probability $P(c)$:
$$P(1ci_n|nrt_{n-1}) + P(2ci_n|nrt_{n-1}) = P(c) = P_c$$
All events which decrements the counter or do not change it are caused with the probability $P(nc)$:
$$P(0cd_n|nrt_{n-1}) + P(cz_n|nrt_{n-1}) = P(nc) = P_{nc}$$
The probability to be in the counter state 1fc or 2fc is the result of incrementing the counter, which happens with the probability $P(c)$:

$$P(1fc_n|nrt_{n-1}) + P(2fc_n|nrt_{n-1}) = P(c) = P_c$$

The probability to be in the counter state 0fc is the result of decrementing / not changing the counter, which happens with the probability $P(nc)$:

$$P(0fc_n|nrt_{n-1}) = P(nc) = P_{nc}$$

or:

$$P(0fc_n|nrt_{n-1}) = P(0cd_n|nrt_{n-1}) + P(cz_n|nrt_{n-1})$$
$$= P(nc)[P(1fc_{n-1}) + P(0fc_{n-1})] / (P(0fc_{n-1}) + P(1fc_{n-1}))$$
$$P(0fc_n|nrt_{n-1}) = P(nc) = P_{nc}$$

with that result we get the probability that the counter reading is 1:

$$P(1ci_n|nrt_{n-1}) = P(1ci_n \cap nrt_{n-1}) / P(nrt_{n-1})$$
$$= P(c)P(0fc_{n-1}) / P(nrt_{n-1})$$
$$P(1ci_n|nrt_{n-1}) = P(c)P(nc) / P(nrt_{n-1}) \qquad (1.2-1)$$

and further, for those who like to continue

$$= P(c)P(nc) / (P(1fc_{n-1}) + P(0fc_{n-1}))$$
$$= P(c)P(nc) / (P(1ci_{n-1}|nrt_{n-2}) + P(nc))$$
$$P(1ci_n|nrt_{n-1}) = P(c) / [1+1/P(nc) \, P(1ci_{n-1}|nrt_{n-2})]$$

and also the probability that the counter reading is 2:

$$1 = P(1ci_n|nrt_{n-1}) + P(2ci_n|nrt_{n-1}) + P(0cd_n|nrt_{n-1}) +$$
$$P(cz_n|nrt_{n-1})$$
$$1 = P(1ci_n|nrt_{n-1}) + P(2ci_n|nrt_{n-1}) + P(nc)$$
$$P(2ci_n|nrt_{n-1}) = P(c) - P(1ci_n|nrt_{n-1})$$

finally we get the probability to trigger the reaction:

$$P(rt_n|nrt_{n-1}) = P(2fc_n|nrt_{n-1})$$
$$= P(2ci_n|nrt_{n-1})$$
$$P(rt_n|nrt_{n-1}) = P(c) - P(1ci_n|nrt_{n-1})$$

and the probability that the reaction is not triggered:

$$P(nrt_n|nrt_{n-1}) = P(1fc_n|nrt_{n-1}) + P(0fc_n|nrt_{n-1})$$
$$= P(nc) + P(1ci_n|nrt_{n-1}) \text{ and using } (1.2-1)$$
$$P(nrt_n|nrt_{n-1}) = P(nc) + P(c)P(nc) / P(nrt_{n-1})$$

1.2.6 P_{nrt} analysis for the 2oo2 code

A graph of the elements of $P(nrt_n)$ shows a clear limit value if P_c is close to 0, but an alternating behavior, when P_c comes close to 1. The single elements come closer and closer to a limit value $P(nrt_\infty|nrt_{\infty-1})$.

The start value for $P(nrt_0) = 1$ as the first event cannot trigger the reaction.

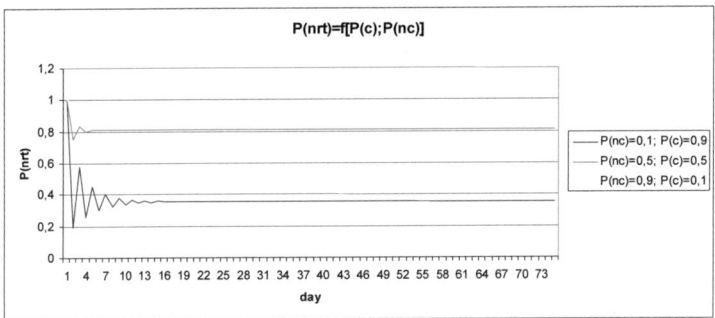

Figure 26: P_{nrt} for the 2oo2 code

This limit value can be seen as the average probability $P(nrt_\infty)$.

To determine this limit value we set two consecutive elements to be equal.

$$P(nrt_n \mid nrt_{n-1}) = P(nrt_{n-1} \mid nrt_{n-2}) = P(nrt_\infty \mid nrt_{\infty-1})$$

$$P(nrt_\infty \mid nrt_{\infty-1}) = P(nc) + \frac{P(c)P(nc)}{P(nrt_\infty \mid nrt_{\infty-1})}$$

$$P(nrt_\infty \mid nrt_{\infty-1})^2 - P(nc)P(nrt_\infty \mid nrt_{\infty-1}) - P(c)P(nc) = 0$$

so we get $P(nrt_\infty)$ by solving this equation with:

$$P(nrt_\infty \mid nrt_{\infty-1}) = \frac{Pnc}{2} + \sqrt{\frac{Pnc^2}{4} + PcPnc}$$

and subsequent ($P_{rt} = 1 - P_{nrt}$):

$$P(rt_\infty \mid nrt_{\infty-1}) = 1 - \frac{Pnc}{2} - \sqrt{\frac{Pnc^2}{4} + PcPnc}$$

For the examples shown in Figure 26 the limit values are:

$P(nc) = 0,1; P(c) = 0,9$: $P(nrt) = 0,3541$

$P(nc) = 0,5; P(c) = 0,5$: $P(nrt) = 0,8090$

$P(nc) = 0,9; P(c) = 0,1$: $P(nrt) = 0,9908$

There are special cases to be mentioned:

$P_c \to 0; \ P_{nc} \to 1$
$P(rt_\infty \mid nrt_{\infty-1}) = 1 - \dfrac{Pnc}{2} - \sqrt{\dfrac{Pnc^2}{4} + PcPnc(Pnc + PcPnc)}$
The extension term $(P_{nc} + P_c P_{nc}) \to 1$ for the used conditions, so it is a multiplication with 1.
$P(rt_\infty \mid nrt_{\infty-1}) \approx 1 - \dfrac{Pnc}{2} - \sqrt{\left(\dfrac{Pnc}{2} + PcPnc\right)^2}$
$P(rt_\infty \mid nrt_{\infty-1}) \approx Pc^2$
$P(nrt_\infty \mid nrt_{\infty-1}) \approx 1 - Pc^2$

$Pnc \to 0; \ Pc \to 1$
$P(rt_\infty \mid nrt_{\infty-1}) = 1 - \dfrac{Pnc}{2} - \dfrac{1}{2}\sqrt{Pnc^2 + 4PcPnc}$
$P(rt_\infty \mid nrt_{\infty-1}) \approx 1 - \sqrt{PcPnc}$
$P(nrt_\infty \mid nrt_{\infty-1}) = \sqrt{PcPnc}$

$P_c = P_{nc} = 0{,}5$
$\begin{aligned} P(rt_\infty \mid nrt_\infty) \ &= 1 - \dfrac{Pnc}{2} - Pnc\sqrt{\dfrac{1}{4} + \dfrac{4}{4}} = 1 - Pnc * \left(\dfrac{1 + \sqrt{5}}{2}\right) \\ &= 1 - Pnc * \varphi = 0{,}19098305... \end{aligned}$
$\begin{aligned} P(nrt_\infty \mid nrt_\infty) \ &= \dfrac{Pnc}{2} + Pnc\sqrt{\dfrac{1}{4} + \dfrac{4}{4}} = Pnc * \left(\dfrac{1 + \sqrt{5}}{2}\right) \\ &= Pnc * \varphi = 0{,}80901695... \end{aligned}$

Where φ is the golden number: 1,6180399…

Table 17: special cases of the 2oo2 code

1.3 2oo3 Analysis

The following transitions of the counter in the 2oo3 code are possible:

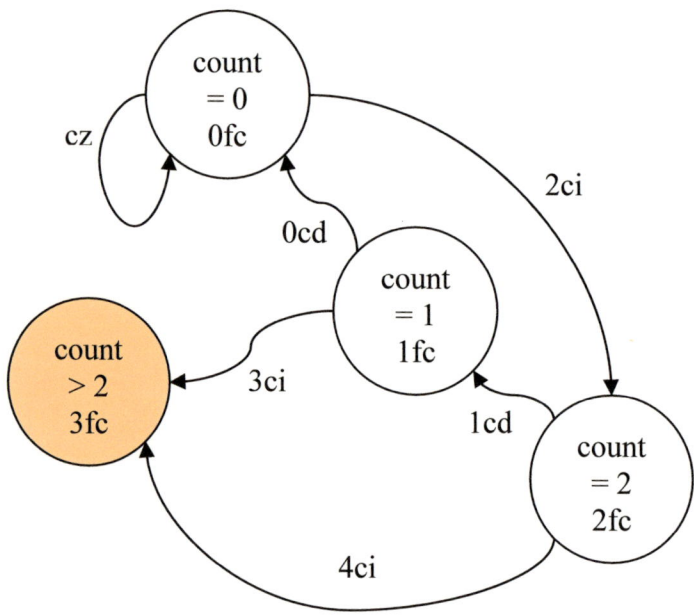

Only in case of the "red" marked state the reaction is triggered by the 2oo3 code.

1.3.1 Probabilities of the transitions and states for the 2oo3 code

The probabilities for the treatment (incrementing / decrementing / not changing) of the counter, the counter states and finally the probabilities to trigger / not to trigger the reaction can be explained, based on the above picture:

1.3.2 Counter changes

The counter will:

- be incremented from 0 to 2, if the event number n is a counting event, with the probability $P(c)$, and the counter reading after event number n-1 was 0:

$$P(2ci_n|nrt_{n-1}) = P(2ci_n \cap nrt_{n-1}) / P(nrt_{n-1})$$
$$= P(c)P(0fc_{n-1}) / (P(0fc_{n-1})+P(1fc_{n-1})+ P(2fc_{n-1}))$$

- be incremented from 1 to 3, if the event number n is a counting event, with the probability $P(c)$, and the counter reading after event number n-1 was 1:

$$P(3ci_n|nrt_{n-1}) = P(3ci_n \cap nrt_{n-1}) / P(nrt_{n-1})$$
$$= P(c)P(1fc_{n-1}) / (P(0fc_{n-1})+P(1fc_{n-1})+ P(2fc_{n-1}))$$

- be incremented from 2 to 4, if the event number n is a counting event, with the probability $P(c)$, and the counter reading after event number n-1 was 2:

$$P(4ci_n|nrt_{n-1}) = P(4ci_n \cap nrt_{n-1}) / P(nrt_{n-1})$$
$$= P(c)P(2fc_{n-1}) / (P(0fc_{n-1})+P(1fc_{n-1})+ P(2fc_{n-1}))$$

- be decremented from 1 to 0, if the event number n is a non-counting event, with the probability $P(nc)$, and the counter reading after event number n-1 was 1:

$$P(0cd_n|nrt_{n-1}) = P(0cd_n \cap nrt_{n-1}) / P(nrt_{n-1})$$
$$= P(nc)P(1fc_{n-1}) / (P(0fc_{n-1})+P(1fc_{n-1})+ P(2fc_{n-1}))$$

- be decremented from 2 to 1, if the event number n is a non-counting event, with the probability $P(nc)$, and the counter reading after event number n-1 was 2:

$$P(1cd_n|nrt_{n-1}) = P(1cd_n \cap nrt_{n-1}) / P(nrt_{n-1})$$
$$= P(nc)P(2fc_{n-1}) / (P(0fc_{n-1})+P(1fc_{n-1})+ P(2fc_{n-1}))$$

- remain 0, if the event number n is a non-counting event, with the probability $P(nc)$, and the counter reading after the event number n-1 was 0:

$$P(cz_n|nrt_{n-1}) = P(cz_n \cap nrt_{n-1}) / P(nrt_{n-1})$$
$$= P(nc)P(0fc_{n-1}) / (P(0fc_{n-1})+P(1fc_{n-1})+ P(2fc_{n-1}))$$

1.3.3 Counter states

The counter reading can:
- only be 2 when the event 2ci occurred:

$$P(2fc_n|nrt_{n-1}) = P(2ci_n|nrt_{n-1})$$

- be >2 when the event 3ci or the event 4ci occurred:

$$P(3fc_n|nrt_{n-1}) = P(3ci_n|nrt_{n-1}) + P(4ci_n|nrt_{n-1})$$

- only be 1 when the event 1cd occurred:

$$P(1fc_n|nrt_{n-1}) = P(1cd_n|nrt_{n-1})$$

- be 0 when the event 0cd or the event cz occurred:

$$P(0fc_n|nrt_{n-1}) = P(0cd_n|nrt_{n-1}) + P(cz_n|nrt_{n-1})$$

1.3.4 Reaction triggered

The reaction is:
- triggered when the counter reading is >2:

$$P(rt_n|nrt_{n-1}) = P(3fc_n|nrt_{n-1})$$

- not triggered when the counter reading is 0, 1 or 2:

$$P(nrt_n|nrt_{n-1}) = P(2fc_n|nrt_{n-1})+P(1fc_n|nrt_{n-1})+ P(0fc_n|nrt_{n-1})$$

1.3.5 Interpretations

The counter will be handled by one of the above actions for sure, this means that the sum of the above probabilities is the almost surely event:

$$P(2ci_n|nrt_{n-1})+P(3ci_n|nrt_{n-1})+P(4ci_n|nrt_{n-1})+$$
$$P(0cd_n|nrt_{n-1})+P(1cd_n|nrt_{n-1})+P(cz_n|nrt_{n-1}) = 1$$

All events which increment the counter are caused with the probability $P(c)$:

$$P(2ci_n|nrt_{n-1})+P(3ci_n|nrt_{n-1})+P(4ci_n|nrt_{n-1}) = P(c) = P_c$$

All events which decrement the counter or do not change it are caused with the probability $P(nc)$:

$$P(1cd_n|nrt_{n-1})+P(0cd_n|nrt_{n-1})+P(cz_n|nrt_{n-1}) = P(nc) = P_{nc}$$

The probability to be in the counter state $2fc$ or $3fc$ is the result of incrementing the counter, which happens with the probability $P(c)$:

$$P(2fc_n|nrt_{n-1})+P(3fc_n|nrt_{n-1}) = P(c) = P_c$$

The probability to be in the counter state $0fc$ or $1fc$ is the result of decrementing / not changing the counter, which happens with the probability $P(nc)$:

$$P(0fc_n|nrt_{n-1})+P(1fc_n|nrt_{n-1}) = P(nc) = P_{nc}$$

Using this formula and taking the formulas from above:

$$P(0fc_n|nrt_{n-1}) = P(0cd_n|nrt_{n-1})+P(cz_n|nrt_{n-1})$$
$$= [\,P(nc)P(1fc_{n-1})+P(nc)P(0fc_{n-1})\,]\,/$$
$$P(nrt_{n-1})$$
$$= P(nc)[P(nc)-P(0fc_{n-1})+P(0fc_{n-1})]\,/$$
$$P(nrt_{n-1})$$

$$P(0fc_n|nrt_{n-1}) = P(nc)^2\,/\,P(nrt_{n-1})$$

This can be used for expressing $P(nrt_n|nrt_{n-1})$:

$$P(nrt_n|nrt_{n-1}) = P(nc)+P(2fc_n|nrt_{n-1})$$
$$= P(nc)+P(2ci_n|nrt_{n-1})$$
$$= P(nc)+P(c)P(0fc_{n-1}|nrt_{n-1})\,/\,P(nrt)_{n-1}$$
$$P(nrt_n|nrt_{n-1}) = P(nc)+P(c)P(nc)^2\,/\,[P(nrt)_{n-1}P(nrt)_{n-2}]$$

1.3.6 P_{nrt} analysis for the 2oo3 code

A graph of the elements of $P(nrt_n)$ shows a clear limit value if P_c is close to 0, but an alternating behavior, when P_c comes close to 1. The single element comes closer and closer to a limit value $P(nrt_\infty|nrt_{\infty-1})$.

The start value for $P(nrt_0) = 1$ as the first event cannot trigger the reaction.

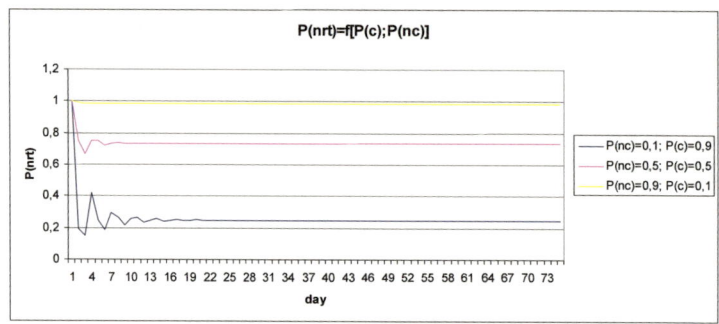

Figure 27: P_{nrt} for the 2oo3 code

This limit value can be seen as the average probability $P(nrt_\infty)$.

To determine this limit value we set three consecutive elements to be equal.

$$P(nrt_n \mid nrt_{n-1}) = P(nrt_{n-1} \mid nrt_{n-2}) = P(nrt_{n-2} \mid nrt_{n-3})$$
$$= P(nrt_\infty \mid nrt_{\infty-1})$$

$$P(nrt_\infty \mid nrt_{\infty-1}) = P(nc) + \frac{P(c)P(nc)^2}{P(nrt_\infty \mid nrt_{\infty-1})P(nrt_\infty \mid nrt_{\infty-1})}$$

$$P(nrt_\infty \mid nrt_{\infty-1})^3 - P(nc)P(nrt_\infty \mid nrt_{\infty-1})^2 - P(c)P(nc)^2 = 0$$

In the following this cubic equation shall be solved by using Cardano's method. But to ease this, the formula is transformed to get rid of P(nc) as factor.

To ease readability we set: $P(nrt_\infty \mid nrt_{\infty-1}) = P_\infty$:

$$P_\infty^3 - P_{nc}P_\infty^2 - P_c P_{nc}^2 = 0 \text{ resp. } P_\infty^3 = P_{nc}P_\infty^2 + P_c P_{nc}^2$$

the transformation is done with $P_\infty = P_{nc} + AP_{nc}$:

$$P_\infty^3 = P_\infty^2 (P_{nc} + AP_{nc}) = P_{nc}P_\infty^2 + P_c P_{nc}^2$$

$$\Rightarrow P_\infty = \sqrt{\frac{P_c P_{nc}}{A}} = P_{nc} + AP_{nc}$$

$$\sqrt{\frac{P_c P_{nc}}{A}} = P_{nc} + AP_{nc} \Rightarrow A^3 + 2A^2 + A - \frac{P_c}{P_{nc}} = 0$$

which is $A^3 + aA^2 + bA + c = 0$ with: $a = 2$, $b = 1$ and

$$c = -\frac{P_c}{P_{nc}}$$

now we apply Cardano's Method:

we reach the equation: $z^3 + pz + q = 0$ with: $z = A + \frac{a}{3}$

and $p = b - \frac{a^2}{3}$, $q = \frac{2a^3}{27} - \frac{ab}{3} + c$

and z is solved by $z = u + v$ where:

$$u = \sqrt[3]{-\frac{q}{2} + \sqrt{D}} \text{ and } v = \sqrt[3]{-\frac{q}{2} - \sqrt{D}} \text{ where}$$

$$D = \left(\frac{q}{2}\right)^2 + \left(\frac{p}{3}\right)^3 \text{ if the condition } uv = -\frac{p}{3} \text{ is true.}$$

we get:
$$p = -\frac{1}{3} \qquad\qquad q = -\frac{2}{27} - \frac{P_c}{P_{nc}}$$

$$D = \frac{1}{4}q^2 - \frac{1}{9^3}$$

$$z = \sqrt[3]{-\frac{q}{2} + \sqrt{D}} + \sqrt[3]{-\frac{q}{2} - \sqrt{D}}$$

z is a function of q, which means it is a function of the relation between P_c and P_{nc}.

the backwards transformations results in $A = z - \frac{2}{3}$:

$$P(nrt_\infty \mid nrt_{\infty-1}) = P_{nc} + AP_{nc} = P_{nc} + P_{nc}\left(z - \frac{2}{3}\right)$$

$$P(nrt_\infty \mid nrt_{\infty-1}) = \frac{1}{3}P_{nc} + P_{nc}z$$

Example ($P_c = 0,2$; $P_{nc} = 0,8$):
$\rightarrow p = -0,3333$, $q = -0,3241$, $z = 0,8463$
$\rightarrow P_{nrt} = 0,9437$

For the examples shown in Figure 27 the limit values are:

$P(nc) = 0,1$; $P(c) = 0,9$: $\qquad$ $P(nrt) = 0,2472$
$P(nc) = 0,5$; $P(c) = 0,5$: $\qquad$ $P(nrt) = 0,7328$
$P(nc) = 0,9$; $P(c) = 0,1$: $\qquad$ $P(nrt) = 0,9837$

1.4 2oo4 Analysis

The following transitions of the counter in the 2oo4 code are possible:

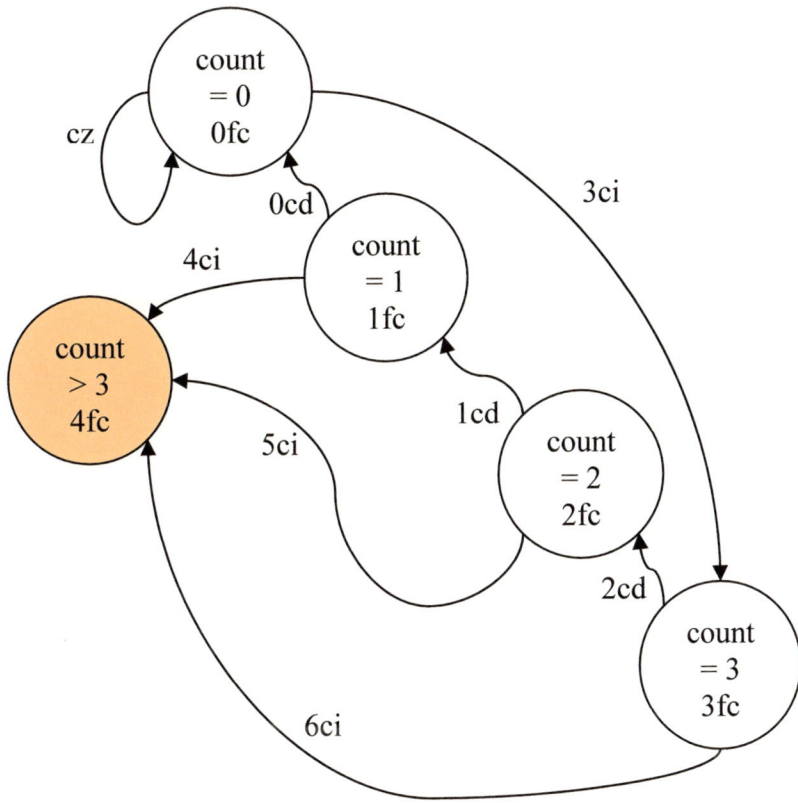

Only in case of the "red" marked state the reaction is triggered by the 2oo4 code.

1.4.1 Probabilities of the transitions and states for the 2oo4 code

The probabilities for the treatment (incrementing / decrementing / not changing) of the counter, the counter states and finally the probabilities to trigger / not to trigger the reaction can be explained, based on the above picture:

1.4.2 Counter changes

The counter will:

- be incremented from 0 to 3, if the event number n is a counting event, with the probability $P(c)$, and the counter reading after event number n-1 was 0:

$$P(3ci_n|nrt_{n-1}) = P(3ci_n \cap nrt_{n-1}) / P(nrt_{n-1})$$
$$= P(c)P(0fc_{n-1}) / (P(0fc_{n-1})+$$
$$P(1fc_{n-1})+P(2fc_{n-1})+P(3fc_{n-1}))$$

- be incremented from 1 to 4, if the event number n is a counting event, with the probability $P(c)$, and the counter reading after event number n-1 was 1:

$$P(4ci_n|nrt_{n-1}) = P(4ci_n \cap nrt_{n-1}) / P(nrt_{n-1})$$
$$= P(c)P(1fc_{n-1}) / (P(0fc_{n-1})+$$
$$P(1fc_{n-1})+P(2fc_{n-1})+P(3fc_{n-1}))$$

- be incremented from 2 to 5, if the event number n is a counting event, with the probability $P(c)$, and the counter reading after event number n-1 was 2:

$$P(5ci_n|nrt_{n-1}) = P(5ci_n \cap nrt_{n-1}) / P(nrt_{n-1})$$
$$= P(c)P(2fc_{n-1}) / (P(0fc_{n-1})+$$
$$P(1fc_{n-1})+P(2fc_{n-1})+P(3fc_{n-1}))$$

- be incremented from 3 to 6, if the event number n is a counting event, with the probability $P(c)$, and the counter reading after event number n-1 was 3:

$$P(6ci_n|nrt_{n-1}) = P(6ci_n \cap nrt_{n-1}) / P(nrt_{n-1})$$
$$= P(c)P(3fc_{n-1}) / (P(0fc_{n-1}) + P(1fc_{n-1}) + P(2fc_{n-1}) + P(3fc_{n-1}))$$

- be decremented from 1 to 0, if the event number n is a non-counting event, with the probability $P(nc)$, and the counter reading after event number n-1 was 1:

$$P(0cd_n|nrt_{n-1}) = P(0cd_n \cap nrt_{n-1}) / P(nrt_{n-1})$$
$$= P(nc)P(1fc_{n-1}) / (P(0fc_{n-1}) + P(1fc_{n-1}) + P(2fc_{n-1}) + P(3fc_{n-1}))$$

- be decremented from 2 to 1, if the event number n is a non-counting event, with the probability $P(nc)$, and the counter reading after event number n-1 was 2:

$$P(1cd_n|nrt_{n-1}) = P(1cd_n \cap nrt_{n-1}) / P(nrt_{n-1})$$
$$= P(nc)P(2fc_{n-1}) / (P(0fc_{n-1}) + P(1fc_{n-1}) + P(2fc_{n-1}) + P(3fc_{n-1}))$$

- be decremented from 3 to 2, if the event number n is a non-counting event, with the probability $P(nc)$, and the counter reading after event number n-1 was 3:

$$P(2cd_n|nrt_{n-1}) = P(2cd_n \cap nrt_{n-1}) / P(nrt_{n-1})$$
$$= P(nc)P(3fc_{n-1}) / (P(0fc_{n-1}) + P(1fc_{n-1}) + P(2fc_{n-1}) + P(3fc_{n-1}))$$

- remain 0, if the event number n is a non-counting event, with the probability $P(nc)$, and the counter reading after event number n-1 was 0:

$$P(cz_n|nrt_{n-1}) = P(cz_n \cap nrt_{n-1}) / P(nrt_{n-1})$$
$$= P(nc)P(0fc_{n-1}) / (P(0fc_{n-1}) + P(1fc_{n-1}) + P(2fc_{n-1}) + P(3fc_{n-1}))$$

1.4.3 Counter states

The counter reading can:
- only be 3 when the event 3ci occurred:

$$P(3fc_n|nrt_{n-1}) = P(3ci_n|nrt_{n-1})$$

- be >3 when the event 4ci or the event 5ci or the event 6ci occurred:

$$P(4fc_n|nrt_{n-1}) = P(4ci_n|nrt_{n-1})+P(5ci_n|nrt_{n-1})+$$
$$P(6ci_n|nrt_{n-1})$$

- only be 2 when the event 2cd occurred:

$$P(2fc_n|nrt_{n-1}) = P(2cd_n|nrt_{n-1})$$

- only be 1 when the event 1cd occurred:

$$P(1fc_n|nrt_{n-1}) = P(1cd_n|nrt_{n-1})$$

- be 0 when the event 0cd or the event cz occurred:

$$P(0fc_n|nrt_{n-1}) = P(0cd_n|nrt_{n-1})+P(cz_n|nrt_{n-1})$$

1.4.4 Reaction triggered

The reaction is:
- triggered when the counter reading is >3:

$$P(rt_n|nrt_{n-1}) \quad = P(4fc_n|nrt_{n-1})$$

- not triggered when the counter reading is 0, 1, 2 or 3:

$$P(nrt_n|nrt_{n-1}) = P(3fc_n|nrt_{n-1})+P(2fc_n|nrt_{n-1})+$$
$$P(1fc_n|nrt_{n-1})+P(0fc_n|nrt_{n-1})$$

1.4.5 Interpretations

The counter will be handled by one of the above actions for sure, this means that the sum of the above probabilities is the almost surely event:

$$P(3ci_n|nrt_{n-1})+P(4ci_n|nrt_{n-1})+P(5ci_n|nrt_{n-1})+$$
$$P(6ci_n|nrt_{n-1})+P(0cd_n|nrt_{n-1})+P(1cd_n|nrt_{n-1})+$$
$$P(2cd_n|nrt_{n-1})+P(cz_n|nrt_{n-1}) = 1$$

All events that increment the fault counter are caused with the probability $P(c)$:

$P(3ci_n|nrt_{n-1})+P(4ci_n|nrt_{n-1})+P(5ci_n|nrt_{n-1})+P(6ci_n|nrt_{n-1})$
$= P(c) = P_c$

All events which decrement the counter or do not change it are caused with the probability $P(nc)$:

$P(0cd_n|nrt_{n-1})+P(1cd_n|nrt_{n-1})+P(2cd_n|nrt_{n-1})+$
$$P(cz_n|nrt_{n-1})$$

$= P(nc) = P_{nc}$

The probability to be in the counter state 3fc or 4fc is the result of incrementing the counter, which happens with the probability $P(c)$:

$P(3fc_n|nrt_{n-1}) +P(4fc_n|nrt_{n-1}) = P(c) = P_c$

The probability to be in the fault counter state 0fc, 1fc or 2fc is the result of decrementing / not changing the counter, which happens with the probability $P(nc)$:

$P(0fc_n|nrt_{n-1}) +P(1fc_n|nrt_{n-1})+P(2fc_n|nrt_{n-1})$
$= P(nc) = P_{nc}$

Using this formula and taking the formulas from above:

$$P(0fc_n|nrt_{n-1}) = P(0cd_n|nrt_{n-1}) +P(cz_n|nrt_{n-1})$$
$$= [\, P(nc)P(1fc_{n-1}) +P(nc)P(0fc_{n-1})\,]\, /$$
$$P(nrt_{n-1})$$
$$= P(nc)[P(nc)-P(0fc_{n-1})-P(2fc_{n-1})+$$
$$P(0fc_{n-1})]\, /\, P(nrt_{n-1})$$
$$= P(nc)^2\, /\, P(nrt_{n-1})-P(nc)P(2fc_{n-1})\, /$$
$$P(nrt_{n-1})$$

with

$$P(2fc_{n-1}|nrt_{n-2}) = P(nc)P(3fc_{n-2})\, /\, P(nrt_{n-2})$$
$$P(0fc_n|nrt_{n-1}) = P(nc)^2\, /\, P(nrt_{n-1})-P(nc)^2P(3fc_{n-2})\, /$$
$$[P(nrt_{n-1})P(nrt_{n-2})]$$

---

with

$$P(3fc_{n-2}|nrt_{n-3}) = P(c)-P(4fc_{n-2}|nrt_{n-3})$$

$$P(0fc_n|nrt_{n-1}) = P(nc)^2 / P(nrt_{n-1})-$$
$$P(c)P(nc)^2 / [P(nrt_{n-1})P(nrt_{n-2})]+$$
$$P(nc)^2 P(4fc_{n-2}|nrt_{n-3}) /$$
$$[P(nrt_{n-1})P(nrt_{n-2})]$$

with

$$P(4fc_{n-2}|nrt_{n-3}) = 1 - P(nrt_{n-2})$$

$$P(0fc_n|nrt_{n-1}) = P(nc)^2 / P(nrt_{n-1})-$$
$$P(c)P(nc)^2 / [P(nrt_{n-1})P(nrt_{n-2})]+$$
$$P(nc)^2 / [P(nrt_{n-1})P(nrt_{n-2})]-$$
$$P(nc)^2 P(nrt_{n-2}) / [P(nrt_{n-1})P(nrt_{n-2})]$$
$$= P(nc)^3 / [P(nrt_{n-1})P(nrt_{n-2})]$$

This can be used for expressing $P(nrt_n|nrt_{n-1})$:

$$P(nrt_n|nrt_{n-1}) = P(3fc_n|nrt_{n-1})+P(2fc_n|nrt_{n-1})+$$
$$P(1fc_n|nrt_{n-1})+P(0fc_n|nrt_{n-1})$$
$$= P(nc)+P(3fc_n|nrt_{n-1})$$
$$= P(nc)P(3ci_n|nrt_{n-1})$$
$$= P(nc)+P(c)P(0fc_{n-1}) / P(nrt_{n-1})$$

$$P(nrt_n|nrt_{n-1}) = P(nc)+P(c)P(nc)^3 /$$
$$[P(nrt)_{n-1}P(nrt)_{n-2} P(nrt)_{n-3}]$$

1.4.6 P_{nrt} analysis for the 2oo4 code

A graph of the elements of $P(nrt_n)$ shows a clear limit value if P_c is close to 0, but an alternating behavior, when P_c comes close to 1. The single element comes closer and closer to a limit value $P(nrt_\infty|nrt_{\infty-1})$.

The start value for $P(nrt_0) = 1$ as the first event cannot trigger the reaction.

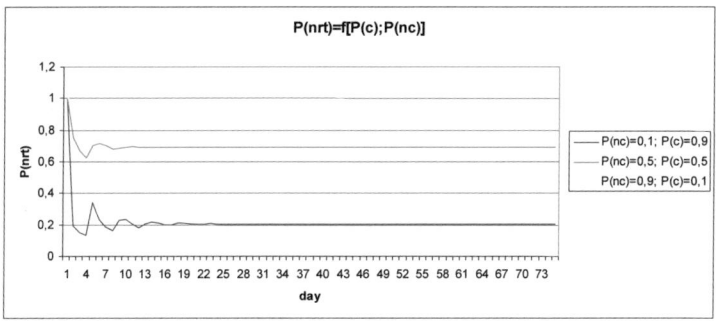

Figure 28: P_{nrt} for the 2oo4 code

This limit value can be seen as the average probability $P(nrt_\infty)$.

To determine this limit value we set four consecutive elements to be equal.

$$P(nrt_\infty \mid nrt_{\infty-1}) \qquad = P(nrt_n \mid nrt_{n-1}) = P(nrt_{n-1} \mid nrt_{n-2})$$
$$= P(nrt_{n-2} \mid nrt_{n-3}) = P(nrt_{n-3} \mid nrt_{n-4})$$

$$P(nrt_\infty \mid nrt_{\infty-1})$$

$$= P(nc) + \frac{P(c)P(nc)^3}{P(nrt_\infty \mid nrt_{\infty-1})P(nrt_\infty \mid nrt_{\infty-1})P(nrt_\infty \mid nrt_{\infty-1})}$$

$$P(nrt_\infty \mid nrt_{\infty-1})^4 - P(nc)P(nrt_\infty \mid nrt_{\infty-1})^3 - P(c)P(nc)^3 = 0$$

In the following this quartic equation shall be solved by using Lodovico Ferrari's method. But to ease this, the formula is transformed to get rid of P(nc) as factor.

To ease readability we set: $P(nrt_\infty \mid nrt_{\infty-1}) = P_\infty$:

$$P_\infty^4 - P_{nc} P_\infty^3 - P_c P_{nc}^3 = 0 \text{ resp. } P_\infty^4 = P_{nc} P_\infty^3 + P_c P_{nc}^3$$

the transformation is done with $P_\infty = P_{nc} + BP_{nc}$:

$$P_\infty^4 = P_\infty^3 (P_{nc} + BP_{nc}) = P_{nc} P_\infty^3 + P_c P_{nc}^3 \Rightarrow P_\infty = \sqrt[3]{\frac{P_c P_{nc}^2}{B}} = P_{nc} + BP_{nc}$$

$$\sqrt[3]{\frac{P_c P_{nc}^2}{B}} = P_{nc} + BP_{nc} \Rightarrow B^4 + 3B^3 + 3B^2 + B - \frac{P_c}{P_{nc}} = 0$$

which is $aB^4 + bB^3 + cB^2 + dB + e = 0$

with $a = 1$, $b = 3$, $c = 3$, $d = 1$ and $e = -\frac{P_c}{P_{nc}}$

now we apply Lodovico Ferrari's Method:
we reach the equation: $u^4 + \alpha u^2 + \beta u + \gamma = 0$ with:

$$u = B + \frac{b}{4a}$$

and $\alpha = -\frac{3b^2}{8a^2} + \frac{c}{a} = -\frac{3}{8}$,

$$\beta = \frac{b^3}{8a^3} - \frac{bc}{2a^2} + \frac{d}{a} = -\frac{1}{8},$$

$$\gamma = -\frac{3b^4}{256a^4} + \frac{b^2 c}{16a^3} - \frac{bd}{4a^2} + \frac{e}{a} = -\frac{3}{256} - \frac{P_c}{P_{nc}}$$

and u is solved by:

$$u = \frac{\pm_1 W \pm_2 \sqrt{-(\alpha + 2y) - 2\left(\alpha \pm_1 \dfrac{\beta}{W}\right)}}{2}$$

The two $\pm_1$ must have the same sign, the $\pm_2$ is independent.

where:

$$W = \sqrt{\alpha + 2y}, \qquad y = \frac{5}{6}\alpha - \frac{P}{3U} + U$$

with $U = \sqrt[3]{-\dfrac{Q}{2} + \sqrt{\dfrac{Q^2}{4} + \dfrac{P^3}{27}}}$, $\qquad P = -\dfrac{\alpha^2}{12} - \gamma = \dfrac{P_c}{P_{nc}}$

and

$$Q = -\frac{\alpha^3}{108} + \frac{\alpha\gamma}{3} - \frac{\beta^2}{8} = \frac{P_c}{8P_{nc}}$$

the backwards transformations results in $B = -\dfrac{b}{4a} + u$:

$$P(nrt_\infty \mid nrt_{\infty-1}) = P_{nc} + BP_{nc} = P_{nc} + P_{nc}\left(-\frac{3}{4} + u\right)$$

$$P(nrt_\infty \mid nrt_{\infty-1}) = \frac{1}{4}P_{nc} + P_{nc}u$$

Example ($P_c = 0,2$; $P_{nc} = 0,8$):
→ $\alpha = -0,375$, $\beta = -0,125$, $\gamma = -0,2617$,
→ $Q = 0,03125$, $P = 0,25$
→ $U = 0,2355$, $y = -0,1941$, $W = 0,1152$, $u = 0,9101$
→ $P_{nrt} = 0,9281$

For the examples shown in Figure 28 the limit values are:

$P(nc) = 0,1$; $P(c) = 0,9$: $P(nrt) = 0,2048$
$P(nc) = 0,5$; $P(c) = 0,5$: $P(nrt) = 0,6901$
$P(nc) = 0,9$; $P(c) = 0,1$: $P(nrt) = 0,9779$

Annex 2

2.1 The probability of the 2oo3 code

This annex shows the different probabilities of the states and transitions of the 2oo3 code for the first three events.

event number	1	2	3
Counter incremented by 2 up to 2	$P(2ci1) = P(c)$	$P(2ci2) = P(c \cap 0fc1)$ $= P(c)P(nc)$	$P(2ci3|nrt2)$ $= P(2ci3 \cap nrt2)/P(nrt2)$ $= P(c)P(0fc2) / P(nrt2)$ $= P(c)P(nc)P(c) / P(nrt2)$
Counter incremented by 2 up to 3	$P(3ci_1) = 0$	$P(3ci_2) = P(c \cap 1fc_1)$ $= 0$	$P(3ci_3|nrt_2)$ $= P(3ci_3 \cap nrt_2)/P(nrt_2)$ $= P(c)P(1fc_2) / P(nrt_2)$ $= P(c)P(nc)P(c) / P(nrt_2)$
Counter incremented by 2 up to 4	$P(4ci_1) = 0$	$P(4ci_2) = P(c \cap 2fc_1)$ $= P(c)P(c)$	$P(4ci_3|nrt_2)$ $= P(4ci_3 \cap nrt_2)/P(nrt_2)$ $= P(c)P(2fc_2) / P(nrt_2)$ $= P(c)P(c)P(nc) / P(nrt_2)$

event number	1	2	3
Counter decremented by 1 down to 0	$P(0cd_1) = 0$	$P(0cd_2) = P(nc \cap 1fc_1)$ $= 0$	$P(0cd_3|nrt_2)$ $= P(0cd_3 \cap nrt_2) / P(nrt_2)$ $= P(nc)P(1fc_2) / P(nrt_2)$ $= P(nc)P(nc)P(c) / P(nrt_2)$
Counter decremented by 1 down to 1	$P(1cd_1) = 0$	$P(1cd_2) = P(nc \cap 2fc_1)$ $= P(nc)P(2fc_1)$ $= P(nc)P(c)$	$P(1cd_3|nrt_2)$ $= P(1cd_3 \cap nrt_2) / P(nrt_2)$ $= P(nc)P(2fc_2) / P(nrt_2)$ $= P(nc)P(c)P(nc) / P(nrt_2)$
Counter remains 0	$P(cz_1) = P(nc)$	$P(cz_2) = P(nc \cap 0fc_1)$ $= P(nc)P(0fc_1)$ $= P(nc)P(nc)$	$P(cz_3|nrt_2)$ $= P(cz_3 \cap nrt_2) /P(nrt_2)$ $= P(nc)P(0fc_2) / P(nrt_2)$ $= P(nc)P(nc)P(nc) / P(nrt_2)$

event number	1	2	3
count is 0	$P(0fc_1) = P(cz_1) = P(nc)$	$P(0fc_2) = P(cz_2)$ $= P(nc)P(nc)$	$P(0fc_3|nrt_2) = P(cz_3 \cup 0cd_3)$ $= P(cz_3|nrt_2)+P(0cd_3|nrt_2)$ $= P(nc)P(nc) / P(nrt_2)$
count is 1	$P(1fc_1) = P(1cd_1) = 0$	$P(1fc_2) = P(1cd_2)$ $= P(nc)P(c)$	$P(1fc_3|nrt_2) = P(1cd_3|nrt_2)$ $= P(nc)P(c)P(nc) / P(nrt_2)$
count is 2	$P(2fc_1) = P(2ci_1) = P(c)$	$P(2fc_2) = P(2ci_2)$ $= P(c)P(nc)$	$P(2fc_3|nrt_2) = P(2ci_3|nrt_2)$ $= P(c)P(nc)P(nc) / P(nrt_2)$
count is ≥3	$P(3fc_1) = P(3ci_1 \cup 4ci_1)$ $= 0$	$P(3fc_2) = P(3ci_2 \cup 4ci_2)$ $= P(3ci_2)+P(4ci_2)$ $= 0 + P(c)P(c)$	$P(3fc_3|nrt_2) = P(3ci_3 \cup 4ci_3)$ $= P(3ci_3|nrt_2)+P(4ci_3|nrt_2)$ $= 2P(c)P(c)P(nc) / P(nrt_2)$

event number	1	2	3
reaction triggered	$P(rt_1) = P(3fc_1) = 0$	$P(rt_2) = P(3fc_2)$ $= P(c)P(c)$	$P(rt_3|nrt_2) = P(3fc_3|nrt_2)$ $= 2P(c)P(c)P(nc) / P(nrt_2)$
reaction not triggered	$P(nrt_1) =$ $P(0fc_1 \cup 1fc_1 \cup 2fc_1)$ $= 1$	$P(nrt_2) =$ $P(0fc_2 \cup 1fc_2 \cup 2fc_2)$ $= P(0fc_2)+P(1fc_2)+$ $P(2fc_2)$ $= 1- P(c)P(c)$	$P(nrt_3|nrt_2) =$ $P(0fc_3 \cup 1fc_3 \cup 2fc_3)$ $= P(0fc_3|nrt_2)+P(1fc_3|nrt_2)+$ $P(2fc_3|nrt_2)$ $= P(nc)P(nc)(1+2P(c)) /$ $P(nrt_2)$

Table 18: The probabilities of the 2oo3 code

After every event the condition $P(rt_x) + P(nrt_x) = 1$ must be fulfilled :

event 1 : $1 = 0 + 1$
event 2 : $1 = P(c) P(c)+1- P(c) P(c)$
event 3 : $1 = 2P(c)P(c) P(nc) / P(nrt2)$ $+ P(nc)P(nc)(1 + 2P(c)) / P(nrt2)$
$= 2P(c)^2 (1 - P(c)) / P(nrt2)$ $+ (1 - P(c))^2(1 + 2P(c)) / P(nrt2)$
$= [2P(c)^2 -2P(c)^3] / P(nrt2)$ $+ (1-2P(c) + P(c)^2)(1 + 2P(c)) / P(nrt2)$
$= [2P(c)^2 -2P(c)^3] / P(nrt2)$ $+$
$\qquad\qquad [1 - 2P(c) + P(c)^2 + 2 P(c) - 4P(c)^2 + 2P(c)^3] / P(nrt2)$
$= [2P(c)^2 -2P(c)^3] / P(nrt2)$ $+ [1 - 3P(c)^2 + 2P(c)^3] / P(nrt2)$
$= [1 - P(c)^2] / P(nrt2)$
$= 1$

Annex 3

3.1 The P_{nrt} for the different codes

$P(nc)$	$P_{nrt\_2ooN} = P(nc) + P(c)\dfrac{P(nc)^{N-1}}{P_{nrt\_2ooN}^{N-1}}$					
$P_{\infty1oo1}$	$P_{\infty2oo2}$	$P_{\infty2oo3}$	$P_{\infty2oo4}$	$P_{\infty2oo5}$	$P_{\infty2oo6}$
0,01	0,104624	0,049846	0,034374	0,027446	0,023580
0,02	0,150357	0,080497	0,058721	0,048452	0,042532
0,03	0,186245	0,106692	0,080397	0,067608	0,060089
0,04	0,216977	0,130371	0,100510	0,085655	0,076798
0,05	0,244374	0,152339	0,119532	0,102919	0,092901
0,06	0,269374	0,173029	0,137724	0,119580	0,108534
0,07	0,292536	0,192709	0,155249	0,135751	0,123784
0,08	0,314226	0,211557	0,172219	0,151511	0,138709
0,09	0,334698	0,229701	0,188712	0,166916	0,153354
0,10	0,354138	0,247237	0,204790	0,182010	0,167751
0,11	0,372687	0,264237	0,220498	0,196825	0,181926
0,12	0,390454	0,280759	0,235874	0,211388	0,195899
0,13	0,407527	0,296851	0,250948	0,225721	0,209686
0,14	0,423977	0,312550	0,265745	0,239842	0,223303
0,15	0,439863	0,327889	0,280285	0,253766	0,236761
0,16	0,455233	0,342894	0,294586	0,267505	0,250069
0,17	0,470130	0,357589	0,308664	0,281072	0,263237
0,18	0,484588	0,371994	0,322532	0,294475	0,276271
0,19	0,498640	0,386126	0,336201	0,307723	0,289179
0,20	0,512311	0,400000	0,349681	0,320823	0,301966
0,21	0,525625	0,413630	0,362980	0,333781	0,314636
0,22	0,538602	0,427028	0,376108	0,346604	0,327195
0,23	0,551263	0,440203	0,389070	0,359297	0,339646
0,24	0,563621	0,453167	0,401874	0,371863	0,351994
0,25	0,575694	0,465927	0,414525	0,384308	0,364241
0,26	0,587493	0,478490	0,427027	0,396635	0,376389
0,27	0,599031	0,490865	0,439386	0,408847	0,388443
0,28	0,610319	0,503057	0,451605	0,420947	0,400403
0,29	0,621366	0,515071	0,463689	0,432938	0,412272
0,30	0,632183	0,526914	0,475641	0,444823	0,424053
0,31	0,642776	0,538590	0,487463	0,456603	0,435746

$P(nc)$	$P_{nrt\_2ooN} = P(nc) + P(c)\dfrac{P(nc)^{N-1}}{P_{nrt\_2ooN}^{N-1}}$				
$P_{\infty 1oo1}$	$P_{\infty 2oo2}$	$P_{\infty 2oo3}$	$P_{\infty 2oo4}$	$P_{\infty 2oo5}$	$P_{\infty 2oo6}$
0,32	0,653153	0,550103	0,499160	0,468281	0,447353
0,33	0,663322	0,561457	0,510732	0,479858	0,458875
0,34	0,673289	0,572656	0,522184	0,491336	0,470315
0,35	0,683060	0,583703	0,533516	0,502718	0,481672
0,36	0,692640	0,594602	0,544731	0,514003	0,492949
0,37	0,702035	0,605355	0,555831	0,525193	0,504145
0,38	0,711249	0,615965	0,566816	0,536289	0,515261
0,39	0,720286	0,626434	0,577690	0,547293	0,526299
0,40	0,729150	0,636764	0,588451	0,558204	0,537258
0,41	0,737846	0,646957	0,599103	0,569024	0,548139
0,42	0,746377	0,657015	0,609646	0,579754	0,558943
0,43	0,754745	0,666940	0,620080	0,590393	0,569670
0,44	0,762955	0,676733	0,630407	0,600942	0,580319
0,45	0,771008	0,686395	0,640627	0,611401	0,590892
0,46	0,778908	0,695928	0,650741	0,621771	0,601387
0,47	0,786657	0,705333	0,660748	0,632052	0,611806
0,48	0,794256	0,714610	0,670651	0,642244	0,622148
0,49	0,801709	0,723761	0,680447	0,652346	0,632413
0,50	0,809017	0,732786	0,690139	0,662359	0,642600
0,51	0,816182	0,741685	0,699725	0,672282	0,652709
0,52	0,823205	0,750459	0,709206	0,682115	0,662739
0,53	0,830088	0,759109	0,718581	0,691858	0,672691
0,54	0,836833	0,767634	0,727851	0,701510	0,682564
0,55	0,843441	0,776035	0,737014	0,711070	0,692356
0,56	0,849912	0,784312	0,746070	0,720538	0,702068
0,57	0,856249	0,792464	0,755020	0,729913	0,711698
0,58	0,862451	0,800491	0,763861	0,739194	0,721245
0,59	0,868520	0,808394	0,772594	0,748381	0,730709
0,60	0,874456	0,816172	0,781217	0,757471	0,740088
0,61	0,880261	0,823824	0,789729	0,766464	0,749380
0,62	0,885934	0,831349	0,798130	0,775360	0,758586
0,63	0,891476	0,838747	0,806418	0,784155	0,767702
0,64	0,896888	0,846017	0,814591	0,792849	0,776728
0,65	0,902170	0,853159	0,822649	0,801439	0,785662
0,66	0,907321	0,860170	0,830589	0,809925	0,794501
0,67	0,912343	0,867050	0,838410	0,818304	0,803244
0,68	0,917235	0,873797	0,846110	0,826574	0,811889

$P(nc)$	$P_{nrt\_2ooN} = P(nc) + P(c)\dfrac{P(nc)^{N-1}}{P_{nrt\_2ooN}^{N-1}}$				
$P_{\infty 1oo1}$	$P_{\infty 2oo2}$	$P_{\infty 2oo3}$	$P_{\infty 2oo4}$	$P_{\infty 2oo5}$	$P_{\infty 2oo6}$
0,69	0,921997	0,880410	0,853687	0,834733	0,820434
0,70	0,926628	0,886888	0,861138	0,842778	0,828875
0,71	0,931129	0,893227	0,868461	0,850706	0,837209
0,72	0,935500	0,899428	0,875653	0,858515	0,845435
0,73	0,939739	0,905487	0,882712	0,866201	0,853548
0,74	0,943847	0,911402	0,889635	0,873761	0,861546
0,75	0,947822	0,917171	0,896417	0,881191	0,869423
0,76	0,951664	0,922791	0,903056	0,888487	0,877177
0,77	0,955373	0,928260	0,909548	0,895645	0,884802
0,78	0,958946	0,933573	0,915888	0,902660	0,892294
0,79	0,962384	0,938728	0,922071	0,909527	0,899647
0,80	0,965685	0,943722	0,928093	0,916240	0,906854
0,81	0,968848	0,948549	0,933948	0,922792	0,913911
0,82	0,971872	0,953207	0,939631	0,929178	0,920808
0,83	0,974754	0,957690	0,945134	0,935388	0,927539
0,84	0,977494	0,961993	0,950451	0,941417	0,934094
0,85	0,980090	0,966111	0,955574	0,947253	0,940464
0,86	0,982540	0,970039	0,960494	0,952887	0,946637
0,87	0,984841	0,973769	0,965202	0,958308	0,952601
0,88	0,986992	0,977296	0,969688	0,963503	0,958342
0,89	0,988990	0,980611	0,973940	0,968457	0,963843
0,90	0,990833	0,983706	0,977944	0,973155	0,969087
0,91	0,992517	0,986572	0,981688	0,977578	0,974053
0,92	0,994041	0,989199	0,985154	0,981705	0,978715
0,93	0,995401	0,991576	0,988324	0,985512	0,983045
0,94	0,996593	0,993691	0,991178	0,988970	0,987009
0,95	0,997614	0,995531	0,993691	0,992047	0,990567
0,96	0,998459	0,997080	0,995835	0,994703	0,993667
0,97	0,999125	0,998322	0,997580	0,996892	0,996250
0,98	0,999608	0,999237	0,998887	0,998554	0,998238
0,99	0,999901	0,999805	0,999711	0,999621	0,999532
0,991	0,999920	0,999842	0,999765	0,999691	0,999619
0,992	0,999937	0,999874	0,999814	0,999755	0,999697
0,993	0,999951	0,999904	0,999857	0,999811	0,999766
0,994	0,999964	0,999929	0,999895	0,999861	0,999827
0,995	0,999975	0,999951	0,999926	0,999903	0,999879
0,996	0,999984	0,999968	0,999953	0,999937	0,999922

$P(nc)$	$P_{nrt\_2ooN} = P(nc) + P(c)\dfrac{P(nc)^{N-1}}{P_{nrt\_2ooN}^{N-1}}$				
$P_{\infty 1oo1}$	$P_{\infty 2oo2}$	$P_{\infty 2oo3}$	$P_{\infty 2oo4}$	$P_{\infty 2oo5}$	$P_{\infty 2oo6}$
0,997	0,999991	0,999982	0,999973	0,999965	0,999956
0,998	0,999996	0,999992	0,999988	0,999984	0,999980
0,999	0,999999	0,999998	0,999997	0,999996	0,999995

Table 19: The probabilities of no reaction of different codes

3.2 The conversion factors for the different codes

$P(nc)$	$x = 1 + \dfrac{P(c)}{P(nc)}\dfrac{1}{x^{N-1}}$				
	N=2	N=3	N=4	N=5	N=6
0,01	10,462430	4,984564	3,437433	2,744629	2,358016
0,02	7,517834	4,024830	2,936034	2,422587	2,126596
0,03	6,208181	3,556401	2,679915	2,253586	2,002964
0,04	5,424429	3,259275	2,512745	2,141385	1,919949
0,05	4,887482	3,046780	2,390636	2,058388	1,858023
0,06	4,489570	2,883822	2,295397	1,992999	1,808905
0,07	4,179092	2,752983	2,217844	1,939300	1,768341
0,08	3,927827	2,644461	2,152732	1,893886	1,733868
0,09	3,718868	2,552235	2,096800	1,854624	1,703936
0,10	3,541381	2,472368	2,047897	1,820096	1,677511
0,11	3,388063	2,402154	2,004527	1,789315	1,653870
0,12	3,253785	2,339662	1,965617	1,761565	1,632488
0,13	3,134826	2,283469	1,930369	1,736314	1,612973
0,14	3,028410	2,232501	1,898176	1,713155	1,595024
0,15	2,932420	2,185925	1,868566	1,691770	1,578407
0,16	2,845208	2,143088	1,841165	1,671908	1,562933
0,17	2,765470	2,103465	1,815673	1,653364	1,548453
0,18	2,692158	2,066633	1,791845	1,635972	1,534841
0,19	2,624419	2,032241	1,769478	1,619594	1,521995
0,20	2,561553	2,000000	1,748403	1,604114	1,509828
0,21	2,502974	1,969667	1,728478	1,589435	1,498267
0,22	2,448193	1,941034	1,709582	1,575475	1,487250

$P(nc)$	$x = 1 + \dfrac{P(c)}{P(nc)}\dfrac{1}{x^{N-1}}$				
	N=2	N=3	N=4	N=5	N=6
0,23	2,396794	1,913928	1,691611	1,562160	1,476724
0,24	2,348423	1,888195	1,674475	1,549431	1,466641
0,25	2,302776	1,863707	1,658098	1,537233	1,456962
0,26	2,259589	1,840347	1,642411	1,525519	1,447651
0,27	2,218634	1,818018	1,627354	1,514247	1,438676
0,28	2,179711	1,796630	1,612875	1,503382	1,430010
0,29	2,142643	1,776107	1,598927	1,492890	1,421628
0,30	2,107275	1,756380	1,585469	1,482742	1,413509
0,31	2,073470	1,737386	1,572462	1,472912	1,405631
0,32	2,041104	1,719071	1,559874	1,463377	1,397977
0,33	2,010067	1,701384	1,547674	1,454115	1,390532
0,34	1,980262	1,684282	1,535835	1,445107	1,383279
0,35	1,951600	1,667724	1,524332	1,436336	1,376207
0,36	1,924001	1,651673	1,513142	1,427785	1,369302
0,37	1,897391	1,636095	1,502246	1,419440	1,362553
0,38	1,871707	1,620960	1,491622	1,411288	1,355950
0,39	1,846886	1,606240	1,481255	1,403315	1,349484
0,40	1,822876	1,591909	1,471129	1,395511	1,343145
0,41	1,799625	1,577943	1,461227	1,387864	1,336925
0,42	1,777087	1,564322	1,451538	1,380366	1,330817
0,43	1,755222	1,551023	1,442047	1,373006	1,324813
0,44	1,733988	1,538029	1,432743	1,365777	1,318907
0,45	1,713352	1,525323	1,423616	1,358670	1,313093
0,46	1,693278	1,512888	1,414654	1,351677	1,307364
0,47	1,673737	1,500709	1,405848	1,344792	1,301716
0,48	1,654701	1,488772	1,397189	1,338008	1,296142
0,49	1,636141	1,477063	1,388668	1,331319	1,290638
0,50	1,618034	1,465571	1,380278	1,324718	1,285199
0,51	1,600356	1,454284	1,372010	1,318200	1,279821
0,52	1,583087	1,443191	1,363857	1,311760	1,274499
0,53	1,566205	1,432281	1,355813	1,305392	1,269229
0,54	1,549691	1,421544	1,347871	1,299092	1,264007
0,55	1,533529	1,410972	1,340025	1,292854	1,258830
0,56	1,517700	1,400556	1,332269	1,286675	1,253693
0,57	1,502191	1,390287	1,324596	1,280549	1,248593
0,58	1,486984	1,380158	1,317002	1,274473	1,243526
0,59	1,472068	1,370160	1,309481	1,268442	1,238490

$P(nc)$	$x = 1 + \dfrac{P(c)}{P(nc)} \dfrac{1}{x^{N-1}}$				
	N=2	N=3	N=4	N=5	N=6
0,60	1,457427	1,360286	1,302028	1,262452	1,233480
0,61	1,443051	1,350531	1,294638	1,256499	1,228493
0,62	1,428926	1,340885	1,287307	1,250580	1,223525
0,63	1,415042	1,331345	1,280029	1,244690	1,218575
0,64	1,401388	1,321902	1,272799	1,238826	1,213638
0,65	1,387954	1,312552	1,265614	1,232984	1,208710
0,66	1,374729	1,303288	1,258469	1,227160	1,203790
0,67	1,361706	1,294104	1,251359	1,221350	1,198872
0,68	1,348875	1,284995	1,244280	1,215551	1,193955
0,69	1,336227	1,275957	1,237227	1,209758	1,189034
0,70	1,323754	1,266982	1,230197	1,203968	1,184107
0,71	1,311450	1,258067	1,223184	1,198178	1,179168
0,72	1,299305	1,249206	1,216185	1,192382	1,174215
0,73	1,287314	1,240393	1,209195	1,186577	1,169244
0,74	1,275468	1,231625	1,202209	1,180758	1,164251
0,75	1,263763	1,222895	1,195223	1,174922	1,159231
0,76	1,252190	1,214199	1,188232	1,169062	1,154180
0,77	1,240744	1,205532	1,181231	1,163176	1,149094
0,78	1,229418	1,196889	1,174215	1,157257	1,143967
0,79	1,218208	1,188264	1,167178	1,151300	1,138793
0,80	1,207107	1,179652	1,160116	1,145299	1,133568
0,81	1,196109	1,171048	1,153022	1,139249	1,128285
0,82	1,185210	1,162447	1,145891	1,133143	1,122937
0,83	1,174403	1,153843	1,138716	1,126974	1,117517
0,84	1,163684	1,145230	1,131489	1,120734	1,112017
0,85	1,153047	1,136602	1,124204	1,114415	1,106428
0,86	1,142488	1,127952	1,116853	1,108008	1,100741
0,87	1,132001	1,119275	1,109428	1,101503	1,094944
0,88	1,121582	1,110564	1,101918	1,094889	1,089025
0,89	1,111225	1,101810	1,094314	1,088154	1,082970
0,90	1,100925	1,093006	1,086605	1,081283	1,076764
0,91	1,090679	1,084145	1,078778	1,074261	1,070388
0,92	1,080480	1,075216	1,070820	1,067070	1,063821
0,93	1,070323	1,066211	1,062714	1,059690	1,057038
0,94	1,060205	1,057118	1,054444	1,052096	1,050010
0,95	1,050120	1,047927	1,045990	1,044260	1,042702
0,96	1,040062	1,038625	1,037328	1,036149	1,035070

$P(nc)$	$x = 1 + \dfrac{P(c)}{P(nc)} \dfrac{1}{x^{N-1}}$				
	N=2	N=3	N=4	N=5	N=6
0,97	1,030026	1,029198	1,028433	1,027723	1,027062
0,98	1,020008	1,019630	1,019272	1,018933	1,018611
0,99	1,010001	1,009904	1,009809	1,009718	1,009628
0,991	1,009001	1,008922	1,008845	1,008770	1,008697
0,992	1,008001	1,007938	1,007877	1,007817	1,007759
0,993	1,007000	1,006952	1,006905	1,006859	1,006814
0,994	1,006000	1,005965	1,005930	1,005896	1,005862
0,995	1,005000	1,004975	1,004951	1,004927	1,004904
0,996	1,004000	1,003984	1,003969	1,003953	1,003938
0,997	1,003000	1,002991	1,002982	1,002974	1,002965
0,998	1,002000	1,001996	1,001992	1,001988	1,001984
0,999	1,001000	1,000999	1,000998	1,000997	1,000996

Table 20: Relation between the different codes